SpringerBriefs in Applied Sciences and Technology

W0079574

More information about this series at http://www.springer.com/series/8884

Carlos O. Maidana

Thermo-Magnetic Systems for Space Nuclear Reactors

An Introduction

 Springer

Carlos O. Maidana
Department of Mechanical Engineering
Chiang Mai University
Chiang Mai
Thailand

ISSN 2191-530X   ISSN 2191-5318   (electronic)
ISBN 978-3-319-09029-0   ISBN 978-3-319-09030-6   (eBook)
DOI 10.1007/978-3-319-09030-6

Library of Congress Control Number: 2014950400

Springer Cham Heidelberg New York Dordrecht London

Printed on acid-free paper

Springer is part of Springer Science+Business Media (www.springer.com)

Acknowledgments

The author wishes to acknowledge and thanks the collaboration and support given by Dr. Hee Reyoung Kim (currently at the Ulsan National Institute of Science and Technology—UNIST), James Werner (Technical Manager/Project Leader at the Idaho National Laboratory—Space Nuclear Systems and Technology Division), Dr. Daniel Wachs (Idaho National Laboratory—Nuclear Fuels and Materials Division, Fuels Performance Department), Jennifer Satten (Editor at Springer Publishing, Physics and Astronomy), to the faculty of the Idaho State University (United States) and Chiang Mai University (Thailand), to Maliwan Boonma, Andrei Chvetsov, Jordi Serrano-Pons, Elaine Kwan and to the indirect contribution of many other people. To all of them, thanks.

Carlos O. Maidana

Contents

About the Author

Dr. Carlos O. Maidana is a Physicist and Research Engineer with 13+ years of experience in research, design and development, as well as in consulting and project management with a vast experience in international and cross disciplinary projects. He is a trade expert, entrepreneur, educator and navy reserve officer off-duty as well. He holds a Ph.D. in Engineering and Applied Science from Idaho State University (USA), a M.Sc. in Physics from Michigan State University (USA), and a B.Sc. in Physics and Applied Physics from the UTN-INSPT (Argentina). He holds a post-doctoral certificate in Space Nuclear Systems Engineering from Washington State University (USA) and the Idaho National Laboratory (USA) as well as several other graduate level certificates (USA, Italy, and Argentina), including studies in research commercialization, government contracting and small business administration. He was born in Buenos Aires, Argentina and resided in countries such as the United States, Switzerland, France and Thailand.

Dr. Maidana was a Research Designer in projects for the U.S. Department of Energy (DoE), the National Aeronautics and Space Administration (NASA), the U.S. Department Homeland Security (DHS), the U.S. Department of Defense, the International Committee for Future Accelerators (ICFA), for the U.S. Advanced Fuel Cycle Initiative (AFCI), for the Measurements and Control Engineering Research Center (MCERC) and a senior research fellow at the European Organization for Nuclear Research (CERN). He has 20+ papers and technical articles published internationally. Dr. Maidana is currently a Lecturer of Mechanical Engineering at Chiang Mai University (Thailand), an Affiliate Research Faculty at Idaho State University (USA) and the founder of a technical consulting and scientific research start-up. Dr. Maidana currently works doing design, modeling, simulation, and optimization of engineering systems and industrial components; in magnetohydrodynamics studies for space nuclear systems and nuclear fusion devices; in the development of compact linear accelerators and associated technology for industrial, security and medical applications; as well as in international cooperation, business development, grant writing, and university level education.

His outreach and community service activities include the development of science, technology and international cooperation at universities in developing countries; advice, counseling and public speaking related to STEM education, career opportunities and curriculum development; He works to educate the public, the media, and government of the benefits of exploring the red planet and outer space advising in questions related to space science and technology with emphasis in space nuclear systems, advanced propulsion systems, and mission planning. Dr. Maidana is a referee for international scientific journals and government organizations as well as for funding bodies besides being a contributor, author, and consultant for the scientific-technical media.

Dr. Maidana is a senior member of the American Institute of Aeronautics and Astronautics (AIAA), serving at the Future and Nuclear Flight Propulsion Technical Committee, and as a scientific officer for the Mars Society Switzerland. He is also a member of the Institute of Electrical and Electronics Engineers (IEEE)—Nuclear and Plasma Sciences Society, the American Physical Society (APS)—Forum of Industrial and Applied Physics, and the Project Management Institute (PMI). He is a registered expert before the European Commission and a CERN external user as well.

Chapter 1
Introduction

Abstract Liquid alloy systems have a high degree of thermal conductivity far superior to ordinary non-metallic liquids. This results in their use for specific heat conducting and dissipation applications. For example, liquid metal-cooled reactors are both moderated and cooled by a liquid metal solution. These reactors are typically very compact and can be used in space propulsion systems and in fission reactors for planetary exploration. Thermo-magnetic systems, such as electromagnetic pumps and electromagnetic flow meters, exploit the fact that liquid metals are conducting fluids capable of carrying currents source of electromagnetic fields useful for pumping and diagnostics. But the coupling between the electromagnetics and thermo-fluid mechanical phenomena gives rise to complex engineering and numerical problems. The environment of operation adds even further complexities with high temperature gradients, radiation and vacuum. This chapter gives an overview of the booklet content, its motivations, describe current and future work and acknowledge people and institutions that without their collaboration this work would not have been possible.

Keywords Liquid metals · Thermomagnetic systems · Space reactors · Space nuclear systems · MHD

Motivation

Liquid alloy systems have a high degree of thermal conductivity far superior to ordinary non-metallic liquids and inherent high densities and electrical conductivities. This results in the use of these materials for specific heat conducting and/or dissipation applications. Typical applications for liquid metals include heat transfer systems, and thermal cooling and heating designs. Uniquely, they can be used to conduct heat and/or electricity between non-metallic and metallic surfaces. The motion of liquid metals in strong magnetic fields generally induces electric currents, which, while interacting with the magnetic field, produce electromagnetic

© The Author(s) 2014   1
C.O. Maidana, *Thermo-Magnetic Systems for Space Nuclear Reactors*,
SpringerBriefs in Applied Sciences and Technology,
DOI 10.1007/978-3-319-09030-6_1

forces. Thermo-magnetic systems, such as electromagnetic pumps or electromagnetic flow meters, exploit the fact that liquid metals are conducting fluids capable of carrying currents source of electromagnetic fields useful for pumping and diagnostics.

Liquid metal-cooled reactors are both moderated and cooled by a liquid metal solution. These reactors are typically very compact and they can be used in regular electric power production, for naval and space propulsion systems or in fission surface power systems for planetary exploration. Liquid metals in fusion reactors can be used in heat exchange, tritium breeder systems and in first wall protection, using a flowing liquid metal surface as a plasma facing component. Many high power particle accelerator facilities will need to employ liquid metal targets and beam dumps for spallation and for heat removal where the severe constraints arising from a megawatt beam deposited on targets and absorbers will require complex procedures to dilute the beam and liquid metals constitute an excellent working fluid due to its intrinsic characteristics. In the metal industry, thermo-magnetic systems are used to transport the molten metal in between processes.

The coupling between the electromagnetics and thermo-fluid mechanical phenomena observed in liquid metal thermo-magnetic systems, and the determination of its geometry and electrical configuration, gives rise to complex engineering magnetohydrodynamics and numerical problems were techniques for global optimization has to be used, MHD instabilities understood -or quantified—and multi-physics models analyzed. The environment of operation adds even further complexity, i.e. vacuum, high temperature gradients and radiation, whilst the presence of external factors, such as the presence of time and space varying magnetic fields, can lead to the need of developing active flow control systems.

This book is an introduction to the design of thermomagnetic systems for liquid metals with emphasis in the design of electromagnetic annular linear induction pumps for space nuclear reactors and tries to fill the gap that exists in the topic from a practical point of view. Concepts on computational physics and mathematical methods are introduced as well as manufacturing and testing procedures. An overview on space nuclear systems is also included. The book is an invaluable tool for design engineers and applied physicists as well as to graduate students in nuclear engineering, mechanical engineering or in applied physics involved with liquid metal technology and space nuclear systems.

Content

An overview on space nuclear systems is given on Chap. 2, where fission surface power systems, radioisotope thermal generators, nuclear thermal propulsion and nuclear electric propulsion are introduced. An introduction to the fundamental equations and concepts used in engineering magnetohydrodynamics is presented on Chap. 3. The reader is briefly presented with the material that has to understand to get involved in this multi-disciplinary field. By no means have we tried to teach or explain a content that would cover hundreds, if not thousands, of pages. The reader should take this chapter as a refresher or a guide of topics to study in specific books fully dedicated to each one of the fields involved. The complexity found in solving engineering problems and analyzing physical phenomena leads to the need of using and developing computational methods and numerical techniques. A brief explanation of the numerical methods used in modeling and simulation is presented in Chap. 4 with the aim of introducing the reader to the nomenclature and methods used in computational engineering sciences. The design methodology of liquid metal electromagnetic pumps of the annular linear induction type, using the electric circuit approach, is developed in Chap. 5. A deeper design analysis and methodology, including the use of first principles and multi-physics analysis, will be available to the interested reader in a future work, which will also include topics such as active flow control, instrumentation and radiation effects on materials and electronic components. Different assembly and fabrication considerations of annular linear induction pumps for space nuclear reactors are provided on Chap. 6. These considerations are the result of the work done by this author at the U.S. Department of Energy's Idaho National Laboratory—Space Nuclear Systems and Technology Division while working on a project for the U.S. National Aeronautics and Space Administration, NASA.

Chapter 2
Overview on Space Nuclear Systems

Abstract Nuclear power sources have enabled or enhanced some of the most challenging and exciting space missions ever conducted. Since 1961, 47 radioisotope thermoelectric generators and 36 space nuclear reactors were successfully flown to provide power for 62 space systems. Yet, the future of nuclear technology for space exploration promises even more remarkable journeys and more amazing discoveries. Space fission nuclear systems can be divided in radioisotope power generators, nuclear thermal propulsion, nuclear electric propulsion and fission surface power technologies. Space radioisotope power systems use radioisotope decay to generate heat and electricity for space missions. For the last fifty-four years, radioisotope thermoelectric generators have provided safe, reliable electric power for space missions where solar power is not feasible. The new advanced sterling radioisotope generators are sought to do an even more efficient job on heat and electricity generation for future space missions. But future space missions will need increased power for propulsion and for surface power applications to support both robotic and human space exploration missions. Nuclear thermal propulsion and nuclear electric propulsion are the most technically mature, advanced propulsion systems that can enable a rapid access to different regions of interest throughout the solar system. The latter is possible by its ability to provide a step increase above what is feasible using a traditional chemical rocket system. Nuclear fission-based power systems are the best suited power sources for surface missions requiring high power in difficult environments where sunlight is limited and reliability is paramount. An overlook of such technologies and activities is presented.

Keywords Space nuclear · Nuclear · Power · Space · RTG · NTP · Nuclear propulsion · Radioisotope

© The Author(s) 2014
C.O. Maidana, *Thermo-Magnetic Systems for Space Nuclear Reactors*,
SpringerBriefs in Applied Sciences and Technology,
DOI 10.1007/978-3-319-09030-6_2

Introduction

The advantage of nuclear power in space applications manifests itself where continuous operation, comparatively high power and the non-dependency of external energy sources is needed. Unlike solar cells, nuclear power systems function independently of sunlight, which is necessary for deep space exploration. Sun intensity reduces as the square of the distance from the Sun and solar panels need to have direct access to sunlight, making them not an alternative for operation during long dark periods. As an example, the solar energy flux on Mars reduces to 50 % compared to that available on Earth and in the Jovian system the solar energy flux reduces to ~0.3 % than that on Earth, Fig. 2.1.

In space applications, it is a great advantage that a small amount of nuclear fuel produces a large amount of energy (low weight-to-capacity ratio). Therefore, nuclear power systems take up much less space than solar power systems and nuclear spacecrafts are easier to orient and direct in space when precision is needed. Furthermore, space nuclear systems can power life support and propulsion systems effectively reducing cost and mission length.

Power sources can be divided in internal and external sources. Among the former, we find chemical and nuclear energy sources whilst among the latter we find solar and beamed power energy as an energy source, Fig. 2.2. Depending on the type of nuclear energy conversion, nuclear power systems can be broadly separated in reactor sources and radioisotope sources. The energy released by reactor power sources are orders of magnitude higher than their radioisotope counter-part.

- Power of reactor nuclear sources is determined by the rate of heavy nuclei fission and can be controlled over a wide range.

Fig. 2.1 Solar energy flux as a function of distance from the Sun in astronomical units (Earth = 1 AU). *Source* NASA

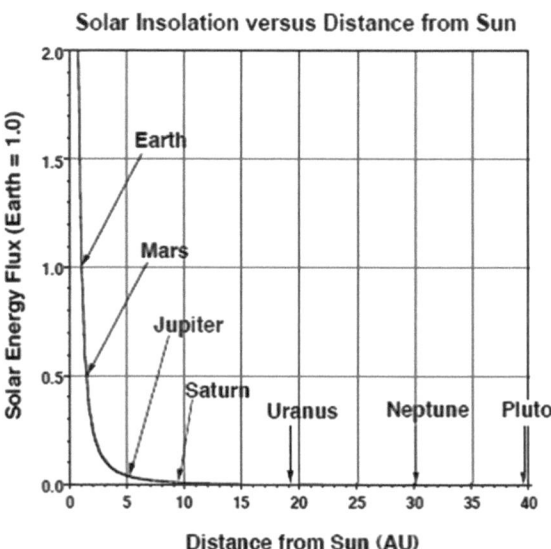

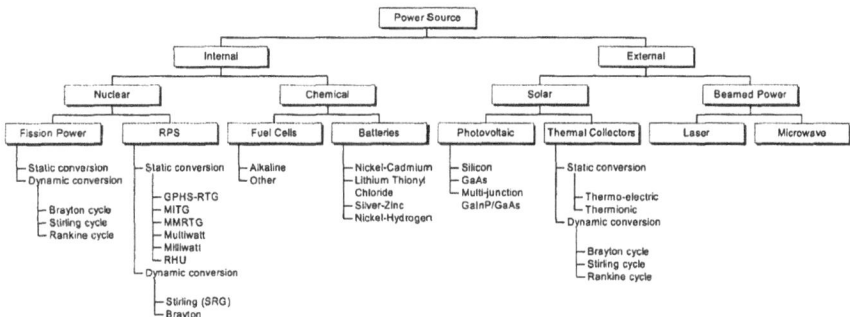

Fig. 2.2 Tree diagram showing internal and external power sources types and their sub-categories

- Thermal power of isotope sources cannot be controlled and it is determined by the type and quantity of radionuclides and decreases in time due to radioactive decay.

Independently of the source, nuclear power is converted into the thermal power that can be absorbed by a working fluid to produce thrust, as in the case of nuclear propulsion systems, or converted into electricity, as in the case of nuclear power sources, through dynamic conversion (i.e. a turbine generator) or by static conversion (i.e. a thermo-ionic convertor or thermo-couples). In general, space fission nuclear systems can be divided in

- radioisotope power generators,
- nuclear thermal propulsion,
- nuclear electric propulsion and
- fission surface power technologies.

Since 1961, the United States has successfully flown 45 radioisotope thermoelectric generators (RTGs) and one reactor to provide power for 25 space systems. The former Soviet Union has reportedly flown at least 35 nuclear reactors and at least two RTGs to power 37 space systems. The latter constitutes a total of 47 radioisotope thermoelectric generators and 36 space nuclear reactors successfully flown to provide power for 62 space systems.

Radioisotope Power Generators

Radioisotope thermoelectric generators (RTGs) are devices that transform the heat produced by a radioactive source into electricity. This is done by the use of an array of thermocouples through Seebeck effect, which is the conversion of temperature differences directly into electricity. Radioisotope heater generators (GHGs) are devices used only to provide heat for the scientific and technical instruments to operate under the low temperatures found in space. Currently RTGs add the functionality of RHGs to their design.

Alpha emitting radioisotopes, having a long half-life compared to the mission lifetime, are generally considered the most attractive nuclear fuels for space applications because of their relatively low shielding requirements and high power densities. Isotopes must not produce significant amounts of gamma, neutron radiation or penetrating radiation in general through other decay modes or decay chain products which would require heavy shielding. Plutonium-238 has the lowest shielding requirements and longest half-life; its power output is 0.54 kW/kg and it needs less than 2.5 mm of shielding, so it has become the most widely used fuel for RTGs, in the form of plutonium (IV) oxide (PuO_2). Its half-life is 87.7 years; it has a reasonable power density, and exceptionally low gamma and neutron radiation levels. Americium-241 is a potential candidate isotope with a longer half-life than ^{238}Pu: 432 years. The power density of ^{241}Am is a quarter that of ^{238}Pu, and because it produces more penetrating radiation through decay chain products than ^{238}Pu, it needs about 18 mm of lead shielding (only ^{238}Pu requires less). RTGs using a material with a half-life λ will diminish in power output by $1 - 0.5^{1/\lambda}$ of their capacity per year. We can also calculate the power decay using the equation: $P_1 = P_0 * 0.9919^Y$ where P_1 is the current power output [W], P_0 is the power output when the RTG was constructed [W] and Y are the years since the RTG was constructed (i.e. If a new RTG outputs 470 W, in 23 years it will output $470 \times 0.83 = 390$ W).

The first radioisotope generator, SNAP-3P, was launched in 1961 having a power of only 2.7 W_e while the latest spacecraft using an RTG, the Mars Science Laboratory, was launched on 2011 providing approximately 2,000 W of thermal power and 120 W of electrical power. Its multi-mission RTG (MMRTG) uses eight general purpose heat source (GPHS) units with a total of 4.8 kg of plutonium oxide. Curiosity, the Mars Science Laboratory rover, uses one RTG to supply heat and electricity for its components and science instruments. A picture of the RTG installed on the NASA's Horizon spacecraft can be seen in Fig. 2.3 while a diagram of general purpose heat source module used in RTGs can be seen in Fig. 2.4. A cut drawing of the GPHS—RTG used for Galileo, Ulysses, Cassini-Huygens and New Horizons space probes is shown in Fig. 2.5.

Stirling radioisotope generators uses free-piston Stirling engines coupled to linear alternators to convert heat to electricity with an average efficiency of 23 %. Greater efficiency can be achieved by increasing the temperature ratio between the hot and cold ends of the generator. Vibration can be eliminated as a concern by implementation of dynamic balancing or use of dual-opposed piston movement. To minimize the risk of the radioactive material being released, the fuel is stored in individual modular units with their own heat shielding. They are surrounded by a layer of iridium metal and encased in high-strength graphite blocks. These two materials are corrosion and heat resistant. Surrounding the graphite blocks is an aero-shell, designed to protect the entire assembly against the heat of re-entering the Earth's atmosphere. The plutonium fuel is also stored in a ceramic form that is heat-resistant, minimising the risk of vaporization and aerosolization. The ceramic is also highly insoluble.

Fig. 2.3 In the clean room at NASA's Kennedy Space Center's Payload Hazardous Servicing Facility, technicians prepare the New Horizons spacecraft for a media event. The RTG seen in this picture is only a mock-up. The real RTG was installed shortly before launch. *Photo* NASA

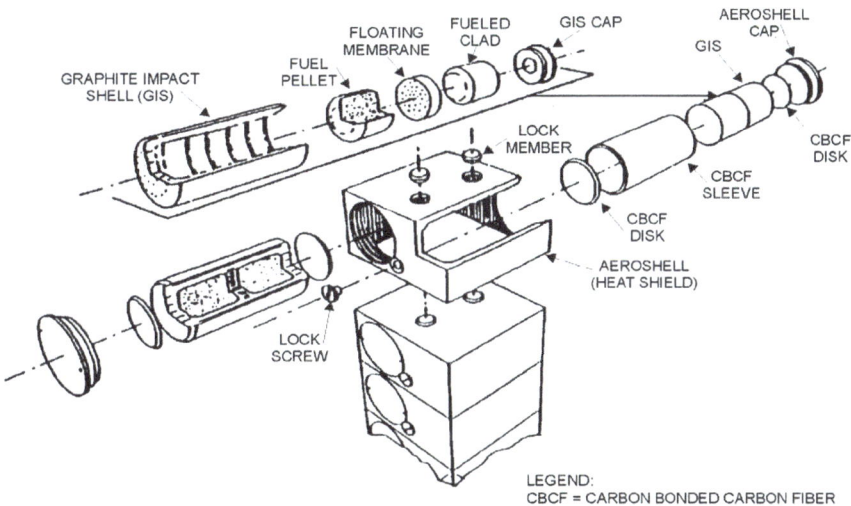

DIAGRAM OF GENERAL PURPOSE HEAT SOURCE MODULE

Fig. 2.4 Diagram of general purpose heat source module used in RTGs. *Courtesy* NASA

GPHS-RTG

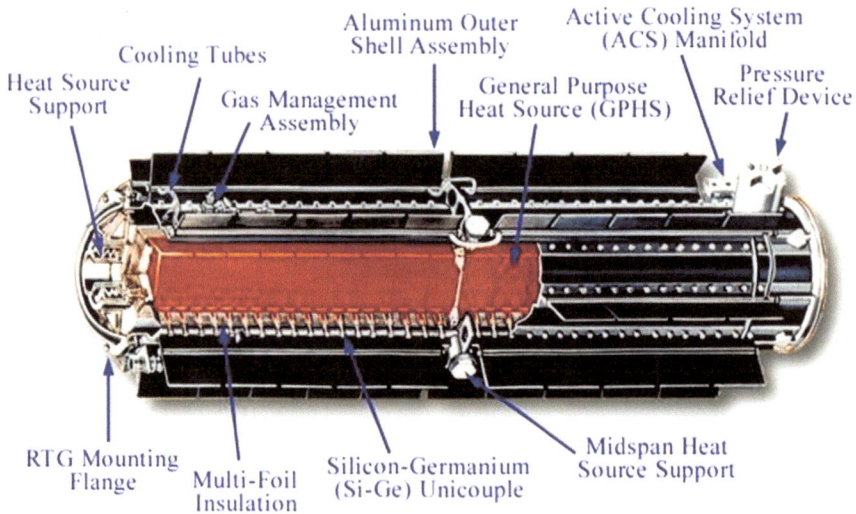

Fig. 2.5 A cut drawing of the general purpose heat source (GPHS)—RTG used for Galileo, Ulysses, Cassini-Huygens and New Horizons space probes

Fission Surface Power

Currently, power is usually generated in space by solar arrays that convert the Sun's energy into electricity or by radioisotope power systems that convert the heat from naturally decaying plutonium-238 into electricity. In order to meet the large power needs for some future missions, solar or radioisotope power systems may be impractical. Space missions will need increased power for propulsion and for surface power applications to support both robotic and human space exploration missions. Nuclear fission reactors are the only reliable power source where sunlight is limited. These reactors are small modular reactors (SMR) with a minimum requirement of 3 kWe and an average of a 30–40 kWe system with 8 or 9 years design life suitable mainly for lunar and Mars surface applications. For the particular case of lunar surface applications an emplaced configuration with regolith shielding augmentation permits near-outpost sitting (<5 rem/year at 100 m separation), Fig. 2.6.

These FSP units should be low temperature, low development risk, liquid-metal cooled reactors and of stainless steel construction. The best options for the nuclear fuel are uranium oxide or uranium metal fuel in stainless steel cladding. Both fuels have extensive fuel performance databases and both fuels bound anticipated performance range for FSP. Uranium oxide is the preferred option mainly for three reasons: it is used as the standard for commercial reactors, the technology is well understood and few commercial entities (none in the United States) are currently

Fig. 2.6 Artistic representation of the NASA/U.S. DoE Fission Surface Power technology unit

making oxide pellet fuel using highly enriched uranium. The flow rate for small 40 kWe FSP units as the one designed by NASA/DoE, Fig. 2.7, is in the order of 80 gpm or 5.03×10^{-3} m^3/s and the operating temperature in the order of 850 K. The working fluids considered for FSP units are Na, Li and NaK78. The disadvantage of NaK78 compared to the other options is the fact that it requires more pumping power than Na or Li but when we study other characteristics of these liquid metals we find a series of advantages that NaK78 has over the other elements

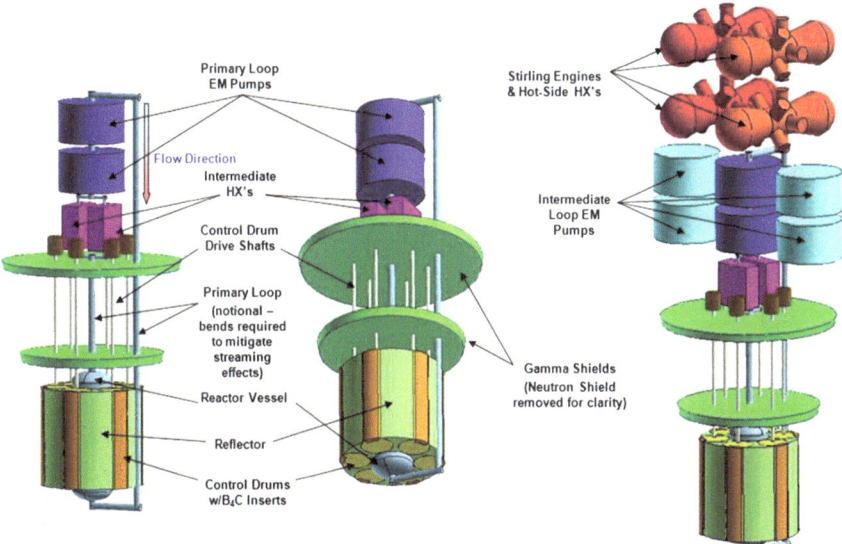

Fig. 2.7 Internal view of the systems that compose the NASA/DoE Fission Surface Power (FSP) technology units

for space applications. Among the characteristics that lead to the selection of NaK78 as the working fluid are:

- NaK78 is liquid at 261 K while Li at 454 K and Na at 361 K;
- Li readily dissolves Nickel at the temperature of interests so it is not suitable to use with steels or super-alloys;
- Neutron capture of ^{6}Li and ^{7}Li produces He gas that would have to be removed from the reactor in space;
- NaK78 activates less strongly than Na.

The heat differential between the 850 K operating temperature and the outside temperature would drive two complementary Stirling engines to turn a 40 kWe generator. Some 100 m^2 of radiators would remove process heat to space.

Nuclear Thermal Propulsion

Chemical rocket propulsion has been the main method of space propulsion: the combustion of a propellant, consisting of fuel and oxidizer components, produces a high speed fluid exhaust that is released by the rocket engines producing thrust. Although this method seems to work extremely well, it has limitations. These limitations make more demanding missions, such as trips to Mars, cumbersome and extremely inefficient. One way to overcome such limitations is the use of nuclear thermal rockets. While in chemical rockets hot gases are created by chemical combustion, in nuclear thermal rockets the hot gas is created by heating the propellant in a nuclear fission reactor.

The ratio of the amount of thrust to the amount of fuel is one key factor when determining a rockets overall effectiveness; this ratio is called the specific impulse. The specific impulse, I_{sp}, is defined as the change in momentum per unit mass of propellant and has SI units of m/s (thrust [Newtons]/mass flow rate [kg/s]) which can be normalized using g_0 (mean acceleration of gravity at Earth's surface) to obtain the engineering units, seconds. In the case of a chemical rocket, the specific impulse is around 500 s while in a nuclear thermal rocket is around 1,000 s. The specific impulse, a measure of engine performance, increases with higher chamber temperatures and lower-weight exhaust gases. Higher specific impulse results in decreased amounts of propellant needed for a given mission.

The basic nuclear thermal rocket engine concept consists of a nuclear reactor heating a low molecular weight gas such as Hydrogen, H_2, to the highest possible temperature, a nozzle for gas expansion and a pump (usually electromagnetic) to make the propellant and cooling circulate through the system. The thrust may be augmented even more by injection of liquid oxygen into the supersonic hydrogen exhaust. In practice, the design becomes quite complex due to the efficiency, weight, temperature and power density requirements. Bimodal versions run electrical systems on board a spacecraft, including radars, as well as provide propulsion.

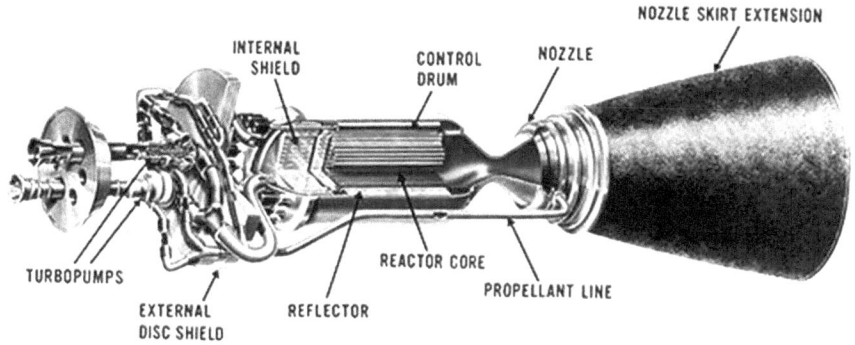

Fig. 2.8 NASA NERVA diagram

Testing on nuclear thermal propulsion (NTP) systems began in the 1960s when the former Soviet Union was searching for a better way to reach space. In the middle of the cold war and in the beginning of the space race, the United States created the program Nuclear Engine for Rocket Vehicle Application (NERVA) in the late 1960s, Fig. 2.8. After a few years of development, the engine was built and tested by NERVA in a remote desert area, with very promising results. NERVA demonstrated that nuclear thermal rocket engines were a feasible and reliable tool for space exploration and at the end of 1968 the U.S. Space Nuclear Propulsion Office certified that the latest NERVA engine, the NRX/XE, met the requirements for a manned Mars mission. The NERVA NRX/EST engine test objectives included:

- Demonstrating engine system operational feasibility
- Showing that no enabling technology issues remained as a barrier to flight engine development.
- Demonstrating completely automatic engine start-up.

The objectives also included testing the use of a new facility for flight engine qualification and acceptance. Total run time was 115 min, including 28 starts. The Rover/NERVA program accumulated 17 h of operating time with 6 h above 2,000 K.

Several other designs have been made during the years, including small nuclear thermal propulsion engines with an approximate weight of 800 kg but political and financial constraints have always limited the development of NTP. The current interest in developing a human mission to Mars made public the inadequacy of pure chemical propulsion systems for space exploration missions to Mars and beyond. The United States space agency, NASA, has partnered with the U.S. Department of Energy and private organizations to continue investigating and to develop nuclear space systems in general and nuclear propulsion systems in particular. This project will test power conversion and thermal management technologies for in-space nuclear power and propulsion systems during the years 2012–2017. Non-nuclear testing will validate the performance of integrated systems. Over the 5-year budget, the project will address five project elements: thermal power conversion, thermal management systems, reactor simulators, thermal propulsion and fission reactors.

Nuclear Electric Propulsion

In nuclear electric systems, nuclear reactors are a heat source for electric ion drives expelling plasma out of a nozzle to propel spacecraft already in space. Magnetic cells ionize hydrogen or xenon, heat it to extremely high temperatures (millions °C), accelerate it and expel it at very high velocity to provide thrust. Nuclear electric propulsion (NEP) is a combination of magneto-plasma-dynamics (MPD) and nuclear power systems. Nuclear power systems are the only alternative for any thruster that consumes more than 100 kW of power. A nuclear electric propulsion system uses a nuclear heat source coupled to an electric generator. The power processing unit converts the electrical power generated by the power source into the power required for each component of the ion thruster. It generates the voltages required by the electromagnetic optics, the discharge chamber and the high currents required for the hollow cathodes. The power management system controls the propellant flow from the propellant tank to the thruster and hollow cathodes. Typically, NEP and NTP can accomplish the same lifting task with similar mass in LEO. When compared to chemical propulsion, NEP was found to accomplish the same missions with 40 % less mass in LEO.

The Variable Specific Impulse Magnetoplasma Rocket (VASIMR) is based on systems for magnetically-confined fusion power for electricity generation, but here the plasma is deliberately leaked to give thrust. The system works most efficiently at low thrust (which can be sustained), with small plasma flow, but high thrust operation is possible. It is very efficient, with 99 % conversion of electric to kinetic energy and specific impulse of 5,000 s. VASIMIR is a design for heavy orbit transfer from Low Earth Orbit (LEO).

Bibliography

Bennett, G.L.: Space nuclear power: opening the final frontier. AIAA paper 2006-4191, presented at the 4th International Energy Conversion Engineering Conference, San Diego, 26–29 June 2006
Houts, M. et al.: Fission Surface Power System Technology for NASA Exploration Missions. National Aeronautics and Space Administration and U.S. Department of Energy internal hand-out
Maidana, C.O. et al.: Design of an annular linear induction pump for nuclear space applications. Proceedings of Nuclear and Emerging Technologies for Space Exploration 2011 (NETS2011), Albuquerque, NM, 7–10 Feb 2011
Werner, J. et al.: An overview of facilities and capabilities to support the development of nuclear thermal propulsion. Proceedings of Nuclear and Emerging Technologies for Space Exploration 2011 (NETS2011), Albuquerque, 7–10 Feb 2011

Chapter 3
Introduction to Engineering Magnetohydrodynamics

Abstract The purpose of this chapter is to give an overview of the basic fundamentals of electromagnetism, fluid mechanics, electric circuits, thermodynamics and heat transfer as they constitute the basis of engineering magnetohydrodynamics. This chapter provides a review of the basic equations needed in the engineering magnetohydrodynamics of liquid metals and plasmas in general and needed to understand how to design thermo-magnetic systems and the processes involved in liquid metal technology.

Keywords MHD · Electromagnetism · Fluid dynamics · Electric circuits · Thermodynamics · Heat transfer · Magnetohydrodynamics

Fundamentals of Electromagnetic Fields and Waves

The electrodynamics of charge particles as well as the existence and propagation of electromagnetic fields and waves is described by Maxwell's equations. They are a set of equations composed by Faraday-Lenz' induction of equation, Maxwell's modified Ampere's equation and Gauss' laws for the electric and magnetic field sources.

$$\nabla \times \boldsymbol{H} = \boldsymbol{J} + \frac{\partial \boldsymbol{D}}{\partial t} \tag{3.1}$$

$$\nabla \times \boldsymbol{E} = -\frac{\partial \boldsymbol{B}}{\partial t} \tag{3.2}$$

$$\nabla \cdot \boldsymbol{D} = \rho \tag{3.3}$$

$$\nabla \cdot \boldsymbol{B} = 0 \tag{3.4}$$

The curl operator is known as vortex density, **H** and **E** are the vectors representing the magnetic and electric field strength, **J** is the current density vector, $\partial \mathbf{D}/\partial t$ is the time derivative of the electric displacement vector **D**, $\partial \mathbf{B}/\partial t$ is the time derivative of the magnetic induction vector **B**, the divergence operator is known as the source

© The Author(s) 2014
C.O. Maidana, *Thermo-Magnetic Systems for Space Nuclear Reactors*,
SpringerBriefs in Applied Sciences and Technology,
DOI 10.1007/978-3-319-09030-6_3

density and ρ is the charge density. The charge ρ and current density $\mathbf{J}$ maybe thought of as the sources of the electromagnetic fields.

It is important to realize that there are two variables to describe the electric properties of the electromagnetic field namely $\mathbf{E}$ and $\mathbf{D}$, and also two variables for the magnetic properties of the field $\mathbf{H}$ and $\mathbf{B}$. This is necessary when certain materials are present with oriented electric and magnetic dipoles. If the electric dipole density is denoted by $\mathbf{P}$ and the magnetic dipole density by $\mathbf{M}$, we can use the following definitions for $\mathbf{D}$ and $\mathbf{B}$:

$$\mathbf{D} = \varepsilon_0 \mathbf{E} + \mathbf{P} = \varepsilon \mathbf{E}, \tag{3.5}$$

$$\mathbf{B} = \mu_0 \mathbf{H} + \mathbf{M} = \mu \mathbf{H} \tag{3.6}$$

where $\varepsilon_0 = 8.854 \times 10^{-12}$ F/m and $\mu_0 = 4\pi \times 10^{-7}$ H/m are the permittivity and the permeability of the vacuum, respectively and ε and μ are the permittivity and the permeability of the material. These relationships are known as constitutive equations. In anisotropic materials, ε depends on the x, y, z direction and the constitutive relations maybe written component-wise in matrix or tensor form:

$$\begin{bmatrix} D_x \\ D_y \\ D_z \end{bmatrix} = \begin{bmatrix} \epsilon_{xx} & \epsilon_{xy} & \epsilon_{xz} \\ \epsilon_{yx} & \epsilon_{yy} & \epsilon_{yz} \\ \epsilon_{zx} & \epsilon_{zy} & \epsilon_{zz} \end{bmatrix} \begin{bmatrix} E_x \\ E_y \\ E_z \end{bmatrix} \tag{3.7}$$

Certain conductors and plasmas in the presence of a constant magnetic field become anisotropic. In non-linear materials, ε may depend on the magnitude of the electric field E,

$$\mathbf{D} = \varepsilon(\mathbf{E})\mathbf{E}$$
$$\varepsilon(\mathbf{E}) = \varepsilon_0 + \varepsilon_1 \mathbf{E} + \varepsilon_2 \mathbf{E}^2 + \cdots . \tag{3.8}$$

A typical consequence of non-linearity is the generation of harmonics. If, for example, $E = E_0 e^{j\omega t}$, then Eq. 3.8 becomes,

$$D = \epsilon(E)E = \epsilon_0 E + \epsilon_1 E^2 + \epsilon_2 E^3 + \cdots$$
$$= \epsilon_0 E e^{j\omega t} + \epsilon_1 E^2 e^{2j\omega t} + \epsilon_2 E^3 e^{3j\omega t} + \cdots \tag{3.9}$$

Thus the input frequency ω generates frequencies 2ω, 3ω, $n^*\omega$...... such non-linear event will then cause the appearance of new frequencies which may be seen as crosstalk in certain RF devices. Materials with a frequency dependence dielectric constant $\varepsilon(\omega)$ are referred to as dispersive. The frequency dependence comes about because when a time-varying electric field is applied, the polarization response of the material cannot be instantaneous. Such dynamic response is described by the convolutional constitutive relationship:

$$\boldsymbol{D}(r,\omega) = \int_{-\infty}^{t} \epsilon\left(t - t'\right) \boldsymbol{E}\left(r, t'\right) dt' \tag{3.10}$$

which becomes multiplicative in the frequency domain:

$$D(r, \omega) = \epsilon(\omega)E(r, \omega) \tag{3.11}$$

All materials are dispersive in certain way but $\varepsilon(\omega)$ usually exhibits a strong dependency on ω for certain frequencies. There are non-linear and dispersive materials that are able to support certain type of non-linear waves called solitons, in which the spreading effect of dispersion is exactly compensated by the non-linearity.

The boundary conditions for the electromagnetic fields across material boundaries is given by,

$$E_{1t} - E_{2t} = 0 \tag{3.12a}$$

$$H_{1t} - H_{2t} = J_s \times \breve{n} \tag{3.12b}$$

$$D_{1n} - D_{2n} = \rho_s \tag{3.12c}$$

$$B_{1n} - B_{2n} = 0 \tag{3.12d}$$

where ρ_s and J_s are the external surface charge and surface current density on the boundary surface (interphase).

The differential form of the charge conservation law is given by,

$$\frac{\partial \rho}{\partial t} + \nabla \cdot J = 0 \tag{3.13}$$

The term on the left is the rate of change of the charge density ρ at a point. The term on the right is the divergence of the current density J. The equation says that the only way for the charge density at a point to change is for a current of charge to flow into or out of the point.

On the other hand, J and E are related by Ohm's law, which represents the third constitutive equation, and can be expressed by,

$$J = \sigma E \tag{3.14}$$

where σ is the conductivity of the material and it is a scalar for isotropic materials and a 3 * 3 matrix (tensor) for anisotropic materials.

The magnetic vector potential, A, is defined in such a way that,

$$\nabla \times A = B \tag{3.15}$$

$$\nabla \cdot A = 0 \tag{3.16}$$

are satisfied. Leading to the operational equation based on Biot-Savart's equation,

$$A = \frac{\mu}{4\pi} \oint \frac{Idl}{R} = \frac{\mu}{4\pi} \iint \frac{J \cdot dSdl}{R} = \frac{\mu}{4\pi} \iiint \frac{JdV}{R} \tag{3.17}$$

The magnetic vector potential can also be defined along with the electric potential by the equations,

$$E = -\nabla V - \frac{\partial A}{\partial t} \quad and \quad B = \nabla \times A \qquad (3.18)$$

Using the 3 constitutive equations and the fact that charge density is zero for propagating electromagnetic waves, we can re-write Maxwell's equations and get the E.M. wave equations.

$$\nabla \times H = (\sigma + j\omega\varepsilon)E \qquad (3.19)$$

$$\nabla \times E = -j\omega\mu H \qquad (3.20)$$

$$\nabla \cdot E = 0 \qquad (3.21)$$

$$\nabla \cdot H = 0 \qquad (3.22)$$

Using the relationship,

$$\nabla \times (\nabla \times F) = \nabla(\nabla \cdot F) - \nabla^2 F \qquad (3.23)$$

on Eqs. 3.19 and 3.20, and operating algebraically, we get the wave equations for the magnetic and electric fields,

$$\nabla^2 H = j\omega\mu(\sigma + j\omega\epsilon)H \equiv \gamma^2 H \qquad (3.24)$$

$$\nabla^2 E = j\omega\mu(\sigma + j\omega\epsilon)E \equiv \gamma^2 E \qquad (3.25)$$

where γ^2 is known as the propagation constant and γ may have real and imaginary components such as,

$$\gamma = \alpha + j\beta \qquad (3.26)$$

For good conductors $\sigma \gg \omega\varepsilon$ and $\alpha = \beta = \mathrm{sqrt}(\pi f\mu\sigma)$ in Eq. 3.26 leading to H to be out of time phase with E at each location by 45° or $\pi/4$ radians. The inverse of α is known as the skin depth, or the depth that the EM wave penetrates a material surface until it decays to $1/e$ of the original signal.

$$\delta = \frac{1}{\alpha} = \frac{1}{\sqrt{\pi f \mu \sigma}} \; (skin\, depth). \qquad (3.27)$$

Fundamentals of Fluid Mechanics

The motion of fluid substances such as liquids and gases are described by a set of differential equations known as the Navier–Stokes equations. These equations state that changes in momentum of fluid particles depend only on the external pressure and internal viscous forces acting on the fluid. The Navier–Stokes equations describe the balance of forces acting at any given region of the fluid. The Navier–Stokes equation can be written as,

$$\rho\left(\frac{D\mathbf{u}}{Dt}\right) = -\nabla p + \nabla \cdot \mathbf{S} + \mathbf{f} \tag{3.28a}$$

Or, instead of writing the total derivative, Eq. 3.28a may be written as

$$\rho\left(\frac{\partial \mathbf{u}}{\partial t} + \mathbf{u} \cdot \nabla \mathbf{u}\right) = -\nabla p + \nabla \cdot \mathbf{S} + \mathbf{f} \tag{3.28b}$$

where ρ is the fluid density, $\mathbf{u}$ the fluid velocity, p the fluid pressure, $\mathbf{S}$ the stress tensor and $\mathbf{f}$ other volumetric forces on the system. This is a set of three scalar equations, one per dimension. By adding other conservation laws and appropriate boundary conditions to the system of equations we get a solvable set of equations. The Navier-Stokes equations are strictly a conservation of momentum statement and don't fully describe the fluid. The extra information needed can be obtained from boundary conditions, i.e. slip, or from relationships stating the conservation of mass, the conservation of energy, or from state equations. The conservation of mass, for example, provides another equation relating the density and the fluid velocity that helps to characterize the fluid,

$$\frac{\partial \rho}{\partial t} + \nabla \cdot \rho \mathbf{u} = 0 \tag{3.29}$$

which for an incompressible fluid becomes,

$$\nabla \cdot \mathbf{u} = 0 \tag{3.30}$$

And the Navier-Stokes equation become,

$$\rho\left(\frac{\partial \mathbf{u}}{\partial t} + \mathbf{u} \cdot \nabla \mathbf{u}\right) = -\nabla p + \mu \nabla^2 \mathbf{u} + \mathbf{f} \tag{3.31}$$

In cylindrical coordinates, the momentum equations, of specific importance for the design of thermo-magnetic systems of annular shape, becomes:

$$
r : \rho\left(\frac{\partial u_r}{\partial t} + u_r\frac{\partial u_r}{\partial r} + \frac{u_\phi}{r}\frac{\partial u_r}{\partial \phi} + u_z\frac{\partial u_r}{\partial z} - \frac{u_\phi^2}{r}\right)
$$
$$
= -\frac{\partial p}{\partial r} + \mu\left[\frac{1}{r}\frac{\partial}{\partial r}\left(r\frac{\partial u_r}{\partial r}\right) + \frac{1}{r^2}\frac{\partial^2 u_r}{\partial \phi^2} + \frac{\partial^2 u_r}{\partial z^2} - \frac{u_r}{r^2} - \frac{2}{r^2}\frac{\partial u_\phi}{\partial \phi}\right] + \rho g_r
$$
$$
\phi : \rho\left(\frac{\partial u_\phi}{\partial t} + u_r\frac{\partial u_\phi}{\partial r} + \frac{u_\phi}{r}\frac{\partial u_\phi}{\partial \phi} + u_z\frac{\partial u_\phi}{\partial z} + \frac{u_r u_\phi}{r}\right)
$$
$$
= -\frac{1}{r}\frac{\partial p}{\partial \phi} + \mu\left[\frac{1}{r}\frac{\partial}{\partial r}\left(r\frac{\partial u_\phi}{\partial r}\right) + \frac{1}{r^2}\frac{\partial^2 u_\phi}{\partial \phi^2} + \frac{\partial^2 u_\phi}{\partial z^2} + \frac{2}{r^2}\frac{\partial u_r}{\partial \phi} - \frac{u_\phi}{r^2}\right] + \rho g_\phi
$$
$$
z : \rho\left(\frac{\partial u_z}{\partial t} + u_r\frac{\partial u_z}{\partial r} + \frac{u_\phi}{r}\frac{\partial u_z}{\partial \phi} + u_z\frac{\partial u_z}{\partial z}\right)
$$
$$
= -\frac{\partial p}{\partial z} + \mu\left[\frac{1}{r}\frac{\partial}{\partial r}\left(r\frac{\partial u_z}{\partial r}\right) + \frac{1}{r^2}\frac{\partial^2 u_z}{\partial \phi^2} + \frac{\partial^2 u_z}{\partial z^2}\right] + \rho g_z. \tag{3.32}
$$

Cylindrical coordinates are chosen to take advantage of symmetry, so that a velocity component can disappear. The equations for axisymmetric flow with the assumption of no azimuthal velocity ($u_\phi = 0$), and the variables independent of ϕ, become:

$$\rho\left(\frac{\partial u_r}{\partial t} + u_r\frac{\partial u_r}{\partial r} + u_z\frac{\partial u_r}{\partial z}\right) = -\frac{\partial p}{\partial r} + \mu\left[\frac{1}{r}\frac{\partial}{\partial r}\left(r\frac{\partial u_r}{\partial r}\right) + \frac{\partial^2 u_r}{\partial z^2} - \frac{u_r}{r^2}\right] + \rho g_r$$

$$\rho\left(\frac{\partial u_z}{\partial t} + u_r\frac{\partial u_z}{\partial r} + u_z\frac{\partial u_z}{\partial z}\right) = -\frac{\partial p}{\partial z} + \mu\left[\frac{1}{r}\frac{\partial}{\partial r}\left(r\frac{\partial u_z}{\partial r}\right) + \frac{\partial^2 u_z}{\partial z^2}\right] + \rho g_z$$

$$\frac{1}{r}\frac{\partial}{\partial r}(ru_r) + \frac{\partial u_z}{\partial z} = 0.$$

$$(3.33)$$

The second term of the left-hand-side of Navier-Stokes equations, as seen in Eq. 3.31, is called convective acceleration or advection term and it could give rise to strong non-linearities. The meaning of the different terms of the Navier-Stoke equation will be examined again from the MHD point of view in another chapter of this booklet.

Solenoid Dynamics

The following discussion applies equally to charge particles and to lagrangian liquid metal particles. It is based on my experience working on particle accelerators and its application on the design of annular linear induction pumps for space nuclear systems at the Idaho National Laboratory on a NASA project.

Solenoids are electromagnetic elements that can be used for focusing along one or more Larmor rotations (long solenoids), the trapping of articles along field lines or point to point focusing (thin solenoids). The solenoids have the big advantage of focusing on both phase space planes at the same time and their higher order components behave in a relatively good way. Their fringe fields propagate over long distances and even though this could be beneficial in certain situations, care must be taken when including them in a lattice. When a charge particle crosses a solenoid fringe fields, it experiences a transverse force causing the particle to follow a spiral path and a coupling between coordinates that can be useful in many situations.

Starting from the relativistic Lagrangian the canonical momentum of a particle of charge q can be expressed as,

$$\vec{P} = \frac{\partial L}{\partial \dot{q}_i} = \gamma m_0\vec{v} + q\vec{A} = \vec{p} + q\vec{A}, \qquad (3.34)$$

where A is the magnetic field vector potential. If the electric field is zero, or nearly zero, in the region where the solenoid is placed, then the canonical momentum is conserved due to the position independence of the Hamiltonian. Now if a particle goes from a region of magnetic field to a region with zero magnetic field, it experiences a change in momentum given by

$$\Delta \vec{P} = q\vec{A} \tag{3.35}$$

In a first approximation where the solenoid only has a constant B_0 magnetic field, its magnetic potential is fully transversal. In this way,

$$\Delta p_\phi = qA_\phi = \frac{1}{2} qr \Delta B_0(z) \tag{3.36}$$

in other words, the particles crossing the fringe fields experience a transverse force (transverse kick in the azimuthal momentum) causing the particles entering (or leaving) with a velocity parallel to the axis of the solenoid to follow a spiral path and a coupling between coordinates. We should highlight again that even though the standard variables are sensitive to edge fields the canonical momentum (and then the canonical variables) are invariant under the effect of these fringe fields. Fringe fields are used as synonymous of edge fields, representing the boundary region between the field and no field region with a quick fall off.

Several software packages allow the use of a variety of cylindrical lenses, in which focusing effects occur only due to fringe field effects. These fringe fields generated by coils extend far longitudinally and contain various nonlinearities due to the longitudinal dependence of the field. In practice it is important to be able to efficiently combine these fields consisting of several solenoidal coils. This could simplify the simulation efforts.

Fundamentals of Thermodynamics and Heat Transfer

As the reader already knows, thermodynamics is the branch of physics and engineering concerned to the study of heat and temperature, and its macroscopic relationship with energy and work. Thermodynamics was mainly developed using empirical postulates and later on explained using the kinetic theory. Its theoretical foundations rely on the physics branch of the statistical mechanics. The properties dealt with in thermodynamics are average properties of the system under study. Heat transfer, on the other hand, is the sub-field of thermodynamics that describes the exchange of thermal energy by heat dissipation.

The thermodynamic state of a system is defined by specifying values of a set of measurable properties sufficient to determine all other properties. The amount of heat transferred in a thermodynamic process that changes the state of a system depends on how that process occurs, not only the net difference between the initial and final states of the process. The fundamental modes of heat transfer are:

- Advection
- Conduction or diffusion

- Convection, and
- Radiation

Advection is the transport mechanism of a fluid substance from one location to another, depending on displacement and its momentum. By transferring matter, energy is moved by the physical transfer of a hot or cold object from one place to another. Conduction is the transfer of energy between objects that are in physical contact. Heat conduction occurs as rapidly moving or vibrating atoms and molecules interact with neighboring particles transferring some of their energy. Convection is the transfer of energy between an object and its environment, due to fluid motion. Convection is essentially the transfer of heat via mass transfer. And radiation is the transfer of energy when the movement of charged particles within atoms is converted to electromagnetic radiation. It is energy emitted by matter as electromagnetic waves.

The main heat transfer mechanism in liquid metal thermo-magnetic systems is conduction. Radiation exists outside the external shell but no convection due to the exposure of the device to the space vacuum. It is very important that the liquid metal not only transport the heat from the nuclear reactor but it also has to absorb enough heat as to keep the device at a controlled temperature.

The heat equation for the temperature variation function $T = T(\mathbf{r}, t)$ is given by:

$$\frac{\partial T}{\partial t} - \alpha \nabla^2 T = 0 \tag{3.37a}$$

for no heat source generation, where α is known as thermal diffusivity and it measures the ability of a material to conduct thermal energy relative to its ability to store thermal energy, $\alpha = \lambda/\rho c_p$, where λ is the thermal conductivity, ρ is the density and c_p the specific heat capacity at constant pressure. In the presence of a heat source, the heat equation becomes:

$$\frac{\partial T}{\partial t} - \alpha \nabla^2 T = \frac{1}{\rho c_p} q_{gen} \tag{3.37b}$$

where q_{gen} is the power generated per unit volume.

Bibliography

Jackson, J.D.: Classical Electrodynamics, 3rd edn. ISBN 0-471-43132-X
Griffiths, D.J.: Introduction to electrodynamics, 3rd edn. Prentice Hall. ISBN 0-13-805326-X
Maidana, C.O.: Design of a Cabinet Safe System for a Portable Particle Accelerator, VDM Verlag (2009), ISBN-10 3639159012, ISBN-13 978-3639159011
Batchelor, G.K.: An Introduction to Fluid Dynamics. Cambridge University Press, Cambridge (1967). ISBN 0-521-66396-2
Landau, L.D., Lifshitz, E.M.: Fluid Mechanics, Course of Theoretical Physics 6, 2nd revised edn. Pergamon Press (1987). ISBN 0-08-033932-8

Chapter 4
Computational Methods

Abstract The complexity found in solving engineering problems and analyzing its physical phenomena leads to the development of computational methods and techniques to find numerical solutions to the set of differential equations describing the process under study. The methods used in computational MHD are mainly a combination of techniques employed in computational fluid dynamics and computational electromagnetism. The complexity arises due to the presence of a magnetic field and its coupling with the fluid. One of the important issues found is to numerically maintain the conservation of magnetic flux condition to avoid any unphysical effects. A brief description of finite elements, finite differences, finite difference time domain, and Monte Carlo methods is presented with the intention of providing a general understanding of the computational and numerical methods used in computational engineering science and computational physics.

Keywords Finite elements · Finite differences · FDTD · Monte Carlo method · Numerical analysis overview

Finite Elements

The Finite Element Method (FEM) is a numerical technique for finding approximate solutions to boundary value problems for differential equations. FEM encompasses all the methods for connecting many simple element equations over many small subdomains, named finite elements, to approximate a more complex equation over a larger domain. Boundary value problems are also called field problems. The field is the domain of interest and most often represents a physical structure. The dependent variables of interest governed by the differential equation are known as field variables and the boundary conditions are the specified values of the field variables (or related variables such as derivatives) on the boundaries of the field. A typical work out of the method involves dividing the domain of the problem into a collection of subdomains, with each subdomain represented

© The Author(s) 2014
C.O. Maidana, *Thermo-Magnetic Systems for Space Nuclear Reactors*,
SpringerBriefs in Applied Sciences and Technology,
DOI 10.1007/978-3-319-09030-6_4

by a set of element equations to the original problem, followed by systematically recombining all sets of element equations into a global system of equations for the final calculation.

While the finite element method is a numerical technique, the Finite Element Analysis (FEA) is a numerical method for solving problems of engineering and mathematical physics. It is useful for problems with complicated geometries, or with physical properties where analytical solutions cannot be obtained. The subdivision of a whole domain into simpler parts has several advantages such as the accurate representation of a complex geometry, the inclusion of dissimilar material properties, the easy representation of the total solution, and the capture of local effects.

Let's assume a two-dimensional case with a single field variable $\varphi(x, y)$ to be determined at every point $P(x, y)$ such that a known governing equation (or equations) is satisfied exactly at every point. Let's also define a node as a specific point in the finite element where the value of the field variable is to be calculated. We should highlight at this point that a finite element is not a differential element of size $dx \times dy$. The values of the field variable computed at the nodes are used to approximate the values at non-nodal points (element interior) by interpolation of the nodal values. For a three-node triangle, the field variable is described by the approximate relation

$$\varphi(x, y) = N_1(x, y)\varphi_1 + N_2(x, y)\varphi_2 + N_3(x, y)\varphi_3$$

where φ_1, φ_2, and φ_3 are the values of the field variable at the nodes, and N_1, N_2, and N_3 are the interpolation functions, also known as shape or blending functions. In the finite element approach, the nodal values of the field variable are treated as unknown constants that are to be determined. The interpolation functions are most often polynomials of the independent variables that satisfy certain required conditions at the nodes. The interpolation functions are predetermined known functions of the independent variables and these functions describe the variation of the field variable within the finite element. The finite element method uses variational methods (calculus of variations) to minimize the error and produce a stable solution.

In general, the number of degrees of freedom associated with a finite element is equal to the product of the number of nodes and the number of values of the field variable (or its derivatives) that must be computed at each node. The primary characteristics of a finite element are embodied in the element stiffness matrix which name originated by its first use in structural mechanics. For a structural finite element, the stiffness matrix contains the geometric and material behavior information that indicates the resistance of the element to deformation when subjected to loading. Such deformation may include axial, bending, shear, and torsional effects. For finite elements used in non-structural analyses, such as fluid flow and heat transfer, the term stiffness matrix is also used since the matrix represents the resistance of the element to change when subjected to external influences.

Example of stiffness matrix: Let's assume a linear elastic spring with spring constant k, Fig. 4.1. As an elastic spring supports axial loading only, we select a

Fig. 4.1 Linear spring with nodes, nodal displacements and forces

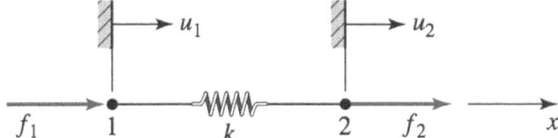

coordinate system oriented along the length of the spring (x axis). Assuming that both the nodal displacements are zero when the spring is undeformed, the net spring deformation is given by $\delta = u_2 - u_1$ and the resultant axial force in the spring is $f = k\delta = k(u_2 - u_1)$. For equilibrium, $f_1 + f_2 = 0$ or $f_1 = -f_2$. Then, in terms of the applied nodal forces,

$$f_1 = -k(u_2 - u_1)$$
$$f_2 = k(u_2 - u_1)$$

which can be expressed in matrix form as

$$\begin{bmatrix} k & -k \\ -k & k \end{bmatrix} \begin{pmatrix} u_1 \\ u_2 \end{pmatrix} = \begin{pmatrix} f_1 \\ f_2 \end{pmatrix}$$

or

$$[k_e]\{u\} = \{f\}$$

where

$$[k_e] = \begin{bmatrix} k & -k \\ -k & k \end{bmatrix}$$

is defined as the element stiffness matrix, $\{u\}$ is the column matrix of nodal displacements, and $\{f\}$ is the column matrix of element nodal forces. The equation shows that the element stiffness matrix for the linear spring element is a 2×2 matrix. This corresponds to the fact that the element exhibits two nodal displacements (degrees of freedom) and that the two displacements are not independent (the body is continuous and elastic). Furthermore, the matrix is symmetric as consequence of the symmetry of the forces (equal and opposite to ensure equilibrium) and the matrix is singular, and therefore not invertible, because the problem as defined as incomplete having no solution: boundary conditions are required.

Finite Differences

Many of the fundamental natural laws governing a wide range of physical phenomena are described by partial derivative equations (PDE). Examples include Laplace's equation, advection-diffusion transport equations, Maxwell's equations of electromagnetism and Navier-Stokes equations for fluid flow among others. Most partial derivative equations of interest in physics and engineering do not have an analytical solution so a numerical procedure must be used to find an approximate solution. The approximation is made at discrete values of the independent variables. In the Finite Difference Method (FDM), all of the derivatives in a differential equation are replaced by algebraic finite difference approximations, which change the differential equation into an algebraic equation that can be solved by simple arithmetic. The finite difference method depends fundamentally on Taylor's theorem and the error between the numerical solution and the exact solution is determined by the error that is committed by going from a differential operator to a difference operator. This error is called the discretization error or truncation error. The term truncation error reflects the fact that a finite part of a Taylor series is used in the approximation.

Assuming a function whose derivatives are to be approximated is properly-behaved, by Taylor's theorem, we can create a Taylor Series expansion such that,

$$f(x_0 + h) = f(x_0) + \frac{f'(x_0)}{1!}h + \frac{f''(x_0)}{2!}h^2 + \cdots + \frac{f^n(x_0)}{n!}h^n + R_n(x)$$

where $n!$ denotes the factorial of n, and $R_n(x)$ is a remainder term, denoting the difference between the Taylor polynomial of degree n and the original function. A first derivative of a function "f" can be obtained by first truncating the Taylor polynomial,

$$f(x_0 + h) = f(x_0) + \frac{f'(x_0)}{1!}h + R_1(x)$$

which after replacing x_0 by a reads,

$$f(a + h) = f(a) + f'(a)h + R_1(x)$$

leading to,

$$f'(a) = \frac{f(a + h) - f(a)}{h} - \frac{R_1(x)}{h}$$

Assuming $R_1(x)$ is sufficiently small, then the first derivative evaluated at the point a can be approximated as,

$$f'(a) \approx \frac{f(a + h) - f(a)}{h}$$

To use a finite difference method to solve a problem, one must first discretize the problem's domain. This is done by dividing the domain into a grid. Meaning that the finite-difference method produces sets of discrete numerical approximations to the derivative, often in a time-stepping manner.

Partial differential equations require proper initial as well as boundary conditions in order to define a well-posed problem. If too many conditions are specified then there will be no solution. If too few conditions are specified then the solution will not be unique.

Finite Difference Time Domain

The finite-difference time-domain (FDTD) method is a numerical analysis technique used for modeling computational electrodynamics problems by finding approximate solutions to the associated system of differential equations. Since it is a time-domain method, FDTD solutions can cover a wide frequency range with a single simulation run, and treat nonlinear material properties in a natural way.

The finite-difference time-domain (FDTD) method is arguably the simplest, both conceptually and in terms of implementation, of the full-wave techniques used to solve problems in electromagnetics. It can accurately tackle a wide range of problems. However, as with all numerical methods, it does have its share of artifacts and the accuracy is contingent upon the implementation. The FDTD method can solve complicated problems, but it is generally computationally expensive. Solutions may require a large amount of memory and computation time. The FDTD method fits into the category of "resonance region" techniques, in which the characteristic dimensions of the domain of interest are somewhere on the order of a wavelength. If an object is very small compared to a wavelength, quasi-static approximations generally provide more efficient solutions. If the wavelength is exceedingly small compared to the physical features of interest, ray-based methods or other techniques may provide a much more efficient way to solve the problem. The time-dependent Maxwell's equations (in partial differential form) are discretized using central-difference approximations to the space and time partial derivatives. The resulting finite-difference equations are solved in a leapfrog manner: the electric field vector components in a volume of space are solved at a given instant in time; then the magnetic field vector components in the same spatial volume are solved at the next instant in time; and the process is repeated over and over again until the desired transient or steady-state electromagnetic field behavior is fully evolved. When Maxwell's differential equations are examined, it can be seen that the change in the E-field in time (time derivative) is dependent on the change in the H-field across space (the curl). This results in the basic FDTD time-stepping relation that, at any point in space, the updated value of the E-field in time is dependent on the stored value of the E-field and the numerical curl of the local distribution of the H-field in space. The H-field is time-stepped in a similar manner. At any point in space, the updated value of the H-field in time is dependent on the stored value of the H-field and the numerical curl of the local distribution of the E-field in space. Iterating the E-field and H-field updates results in a marching-in-time process wherein sampled-data analogs of the continuous electromagnetic waves under consideration propagate in a numerical grid stored in the computer memory.

Consider the Taylor series expansions of the function $f(x)$ expanded about the point x_0 with an offset of $\pm\delta/2$:

$$f\left(x_0 + \frac{\delta}{2}\right) = f(x_0) + \frac{\delta}{2}f'(x_0) + \frac{1}{2!}\left(\frac{\delta}{2}\right)^2 f''(x_0) + \frac{1}{3!}\left(\frac{\delta}{2}\right)^3 f'''(x_0) + \cdots$$

$$f\left(x_0 - \frac{\delta}{2}\right) = f(x_0) - \frac{\delta}{2}f'(x_0) + \frac{1}{2!}\left(\frac{\delta}{2}\right)^2 f''(x_0) - \frac{1}{3!}\left(\frac{\delta}{2}\right)^3 f'''(x_0) + \cdots$$

Subtracting the second equation from the first yields,

$$f\left(x_0 + \frac{\delta}{2}\right) - f\left(x_0 - \frac{\delta}{2}\right) = \delta f'(x_0) + \frac{2}{3!}\left(\frac{\delta}{2}\right)^3 f'''(x_0) + \cdots$$

$$\frac{f\left(x_0 + \frac{\delta}{2}\right) - f\left(x_0 - \frac{\delta}{2}\right)}{\delta} = f'(x_0) + \frac{1}{3!}\left(\frac{\delta}{2}\right)^2 f'''(x_0) + \cdots$$

Thus the term on the left is equal to the derivative of the function at the point x_0 plus a term which depends on δ^2 plus an infinite number of other terms which are not shown. For the terms which are not shown, the next would depend on δ^4. If δ is sufficiently small, a reasonable approximation to the derivative may be obtained and the central-difference approximation is given by,

$$\frac{df(x)}{dx}|_{x=x_0} = \frac{f\left(x_0 + \frac{\delta}{2}\right) - f\left(x_0 - \frac{\delta}{2}\right)}{\delta}$$

The central difference provides an approximation of the derivative of the function at x_0, but the function is not actually sampled there. Instead, the function is sampled at the neighboring points $x_0 + \delta/2$ and $x_0 - \delta/2$. Since the lowest power of δ being ignored is second order, the central difference is said to have second-order accuracy or second-order behavior. If δ is reduced by a factor of 10, the error in the approximation should be reduced by a factor of 100.

The FDTD is usually implemented using the Yee algorithm. The Yee algorithm can be summarized as follows:

1. Replace all the derivatives in Ampere's and Faraday's laws with finite differences. Discretize space and time so that the electric and magnetic fields are staggered in both space and time.
2. Solve the resulting difference equations to obtain updated equations to express the (unknown) future fields in terms of (known) past fields.
3. Evaluate the magnetic fields one time-step into the future so they are now known (effectively they become past fields).
4. Evaluate the electric fields one time-step into the future so they are now known (effectively they become past fields).
5. Repeat the previous two steps until the fields have been obtained over the desired duration.

To implement an FDTD solution of Maxwell's equations, a computational domain must first be established. The E and H fields are determined at every point in space

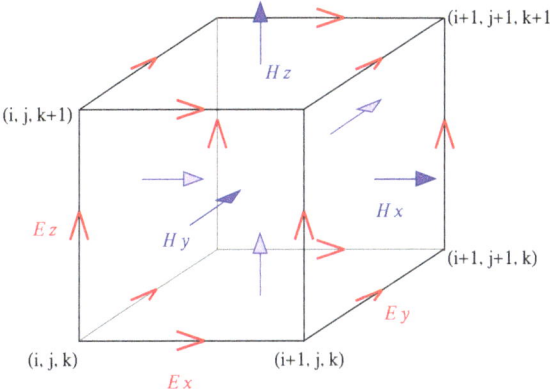

Fig. 4.2 Illustration of a standard Cartesian Yee cell used for FDTD, about which electric and magnetic field vector components are distributed. Visualized as a cubic voxel, the electric field components form the edges of the cube, and the magnetic field components form the normal to the faces of the cube. A three-dimensional space lattice consists of a multiplicity of such Yee cells. An electromagnetic wave interaction structure is mapped into the space lattice by assigning appropriate values of permittivity to each electric field component, and permeability to each magnetic field component. *Courtesy* Steven G. Johnson

within that computational domain and the material of each cell within the computational domain must be specified. Once the computational domain and the grid materials are established, a source is specified. The source can be current on a wire, applied electric field or impinging plane wave. Figure 4.2 shows a Yee lattice picture, illustrating the staggered grid used for the finite-difference time-domain method in electromagnetism.

Monte Carlo Method

The Monte Carlo method is a means of statistical evaluation of mathematical functions using random samples and it is also a method for exploring the sensitivity of a complex system by varying parameters within statistical constraints. Because of their reliance on repeated computation of random or pseudo-random numbers, these methods are most suited to calculation by a computer and tend to be used when it is unfeasible or impossible to compute an exact result with a deterministic algorithm.

For example, let's consider a circle inscribed in a unit square. Given that the circle and the square have a ratio of areas that is $\pi/4$, the value of π can be approximated using a Monte Carlo method:

1. Draw a square on the ground, and then inscribe a circle within it.
2. Uniformly scatter small objects of uniform size over the square.

3. Count the number of objects inside the circle, m, and the total number of objects, n.
4. The ratio of the two counts is an estimate of the ratio of the two areas, which is $\pi/4$. Multiply the result by 4 to estimate π. ($\pi = 4m/n$)

If the objects are not uniformly distributed, then the approximation will be poor. There should be a large number of inputs. The accuracy of the method is directly proportional to the number of samples (bins) used.

Optimization

Optimization is the selection of a best element (with regard to some criteria) from some set of available alternatives and engineering or design optimization is the subject which uses optimization techniques to achieve design goals in engineering. Optimization includes finding best available values of some objective function given a defined domain or a set of constraints including a variety of different types of objective functions and different types of domains. And a more formal definition would be: "Optimization is a subject that deals with the problem of minimizing or maximizing a certain function in a finite dimensional Euclidean space over a subset of that space, which is usually determined by functional inequalities".

Adding more than one objective to an optimization problem adds complexity. The set of trade-off designs that cannot be improved upon according to one criterion without hurting another criterion is known as the *Pareto set*. A design is judged to be *Pareto optimal*, or in the *Pareto set*, if it is not dominated by any other design: If it is worse than another design in some respects and no better in any respect, then it is dominated and is not Pareto optimal. Optimization problems are often multi-modal; that is, they possess multiple good solutions. They could all be globally good or there could be a mix of globally good and locally good solutions. Obtaining all (or at least some of) the multiple solutions is the goal of a multi-modal optimizer. Classical optimization techniques due to their iterative approach do not perform satisfactorily when they are used to obtain multiple solutions, since it is not guaranteed that different solutions will be obtained even with different starting points in multiple runs of the algorithm. Evolutionary Algorithms are however a very popular approach to obtain multiple solutions in a multi-modal optimization task.

The formulation of an optimization problem begins with identifying the design variables to be varied during the optimization process. A design problem usually involves many design parameters, of which some are highly sensitive to the proper working of the design. These parameters are called design variables. Other not so important design parameters usually remain fixed or are quasi-dependent of the design variables. In most engineering problems, there could be more than one objective that the designer might want to optimize simultaneously. The multiple objective optimization algorithms are usually complex and computationally expensive.

Bibliography

Yee, K.: Numerical solution of initial boundary value problems involving Maxwell's equations in isotropic media. IEEE Trans. Antennas Propag. **14**(3), 302–307 (1966)

Schneider, J.: Introduction to the finite-difference time-domain method lecture notes (2013)

Rao, S.S.: Engineering Optimization: Theory and Practice, Wiley, New York (2009)

Hildebrand, F.B.: Finite-Difference Equations and Simulations. Prentice-Hall, Englewood Cliffs, New Jersey (1968)

Chapter 5
Design of Annular Linear Induction Pumps for Space Nuclear Reactors

Abstract In space reactors as well as in other types of semi-transportable small modular reactors, weight, reliability and efficiency are of fundamental importance. Furthermore, for space reactors liquid metals as working fluids are the only option due to the working environment characteristics that outer space provides. For space power systems, the induction electromagnetic pump is inherently more reliable than the conduction electromagnetic pump. The annular linear induction pump also has several advantages over its flat counterpart because it has greater structural integrity, it is more adaptable to normal piping systems, and it allows greater design freedom in the coil configuration. The annular design also has a basically greater output capability since the path followed by the induced currents has a lower resistance than the path followed in a corresponding flat pump. The design work of ALIP-type thermo-magnetic systems using the electric circuits approach is described as well as the fundamental ideas behind the development of modelling and simulation techniques for the development of an optimized design methodology using "first principles". This project, involves the use of theoretical, computational and experimental tools for multi-physics analysis as well as advanced engineering design methods and techniques.

Keywords Electromagnetic pump · Annular linear induction pump · Thermo-magnetic systems · Liquid metal · Engineering MHD · Magnetohydrodynamics · Space reactors · Fission surface power · Space nuclear systems · FSP

MHD Equations and Working Fundamentals

Annular Linear Induction Pumps make use of a 3-phase alternating current (ac) input voltage to power a group of solenoids that generates a traveling magnetic field along the pump duct, Fig. 5.1. This traveling magnetic field induces a current on the surface of the liquid metal in the duct and as a result, an electromagnetic (EM) force is generated. This electromagnetic force will be the one pumping the molten metal through the duct.

© The Author(s) 2014   33
C.O. Maidana, *Thermo-Magnetic Systems for Space Nuclear Reactors*,
SpringerBriefs in Applied Sciences and Technology,
DOI 10.1007/978-3-319-09030-6_5

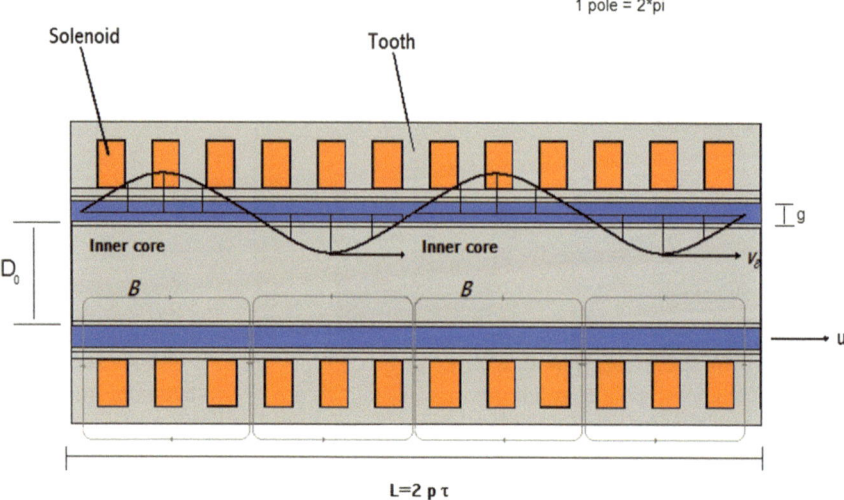

Fig. 5.1 Section of an annular linear induction pump showing the inner core, duct, solenoids, teeth and traveling magnetic wave

One way to model the current, I, of the primary windings is by an equivalent current sheet at the outer wall of the pump:

$$J_s(z,t) = J_a e^{i(\omega t - kz)} \tag{5.1}$$

where $k = k/\tau$, τ is the pole pitch, ω is the angular frequency, $J_a = 3\sqrt{2}kNI/p\tau$, p the number of pole pairs and N is the number of turns per slot. This is a sinusoidal wave with wavelength L and speed, or synchronous velocity, $v_b = \omega L/2\pi$, the pole pitch being $\tau = L/2$.

The equations describing the pumping process in the duct are:

$$J_i = \sigma(\mathbf{E} + \mathbf{u} \times \mathbf{B}) \tag{5.2}$$

$$\rho\left[\frac{\partial \mathbf{u}}{\partial t} + (u \cdot \nabla)\mathbf{u}\right] + \nabla p - \rho v \nabla^2 \mathbf{u} = \mathbf{J} \times \mathbf{B} \tag{5.3}$$

where the current density is $\mathbf{J} = \mathbf{J}_s + \mathbf{J}_i$, σ and v are the electric conductivity and kinematic viscosity (ratio of the viscous force to the inertial force) of the fluid, and **u** is the fluid velocity. Assuming a steady state condition, one can get:

$$\frac{F_z}{2\pi Rg} = \frac{12\rho v \langle u \rangle}{g^2} + \Delta p \tag{5.4}$$

where F_z is the Lorentz force generated, Δp is the total pressure developed, R the mean radius of the fluid and g is the dimension of the annular gap. It is possible to use the vector potential **A** instead of the magnetic field **B** for certain calculations; for the latter, possible gauges are $\nabla \cdot \mathbf{A} = 0$ (coulomb gauge) or $\nabla \cdot \mathbf{A} + \mu\sigma\varphi = 0$, where A is defined as $\nabla \times \boldsymbol{A} = \mathbf{B}$, μ is the magnetic permeability and ϕ the electric potential.

Because the linear momentum of the fluid element could change not only by the pressure force, $-\nabla p$, viscous friction, $\rho v \nabla^2 \vec{u}$, and Lorentz force, $\vec{J} \times \vec{B}$, but also by volumetric forces of non-electromagnetic origin, Eq. (5.3) should be modified and it could be expressed with an additional term f in the right hand side,

$$\rho \left[\frac{\partial \mathbf{u}}{\partial t} + (u \cdot \nabla)\mathbf{u} \right] + \nabla p - \rho v \nabla^2 \mathbf{u} - \mathbf{f} = \mathbf{J} \times \mathbf{B} \qquad (5.5)$$

while the conservation of mass for liquid metals would be given by $\nabla \cdot \vec{u} = \mathbf{0}$, which expresses the incompressibility of the fluid.

An induction equation, valid in the domain occupied by the fluid and generated by the mechanical stretching of the field lines due to the velocity field, can be written as,

$$\frac{\partial}{\partial t}\boldsymbol{B} + (\boldsymbol{u} \cdot \nabla)\boldsymbol{B} = \frac{1}{\mu\sigma}\nabla^2\boldsymbol{B} + (\boldsymbol{B} \cdot \nabla)\boldsymbol{u} \qquad (5.6)$$

describing the time evolution of the B field, $\partial\boldsymbol{B}/\partial t$, due to advection $(\boldsymbol{u} \cdot \nabla)\boldsymbol{B}$, diffusion $\nabla^2\boldsymbol{B}$ and field intensity sources $(\boldsymbol{B} \cdot \nabla)\boldsymbol{u}$. The term $(\mu\sigma)^{-1}$ is also known as the magnetic diffusivity.

One of the solutions of the induction equation gives rise to the possibility of self-excitation of the $\boldsymbol{B}$ field even with no external field sources due to small perturbations once the production of magnetic intensities by mechanical stretching overcomes their damping by diffusion. This partially explains certain effects found in EM pumps during the start-up and shut down periods. Sometimes the induction equation, Eq. (5.6), is written dimensionless by the introduction of scale variables and as a function of what is called the magnetic Reynolds number, $R_m = \mu\sigma L u_0$, where u_0 is the mean velocity and L the characteristic length. A relatively small R_m generates only small perturbations on the applied field; if R_m is relatively large then a small current creates a large induced B field. For small magnetic Reynolds numbers ($R_m \ll 1$), the magnetic field will be dominated by diffusion and perturbative methods can be used accurately. Expressing the magnetic Reynolds number as $R_m = (g/\delta_m K_c) \cdot (\sigma\mu\omega/k^2) \cdot s$ where δ_m is the wall thickness (non-magnetic gap), K_c is the Carter's coefficient and s the slip factor, it was found experimentally that instabilities arises when the following three conditions hold true:

- $R_{ms} > 1$,
- the ratio of the mean radius to the pole pitch (D/2τ), and
- the modified interaction parameter N_{int} are large enough.

The modified interaction parameter is expressed as $N_{int} = 2gB_{ar}^2 \sigma/\rho\lambda v_b$ where B_{ar} is the radial component of the applied magnetic field and λ the friction coefficient. This specific instability appears as a low frequency pulsation in the pressure head affecting the flow rate, liquid metal velocity and magnetic field distribution. As a consequence vortices are generated in the inlet region as well as fluctuations in the winding currents and voltages. Araseki et al. have reported that the dominant frequency of this instability is in the range 0–10 Hz with amplitude that increases with the slip factor.

In a similar way, one can find that the equation for temperature is

$$\rho c_p \left[\frac{\partial}{\partial t} T + (\boldsymbol{u} \cdot \nabla) T \right] = \nabla \cdot (\lambda \nabla T) + \frac{1}{\sigma} J^2 + \Phi + Q \tag{5.7}$$

which is a convection-diffusion equation with λ: thermal conductivity, Q: other sources of volumetric energy release such as radiation or chemical reactions and thermal diffusivity : λ/c_p, c_p: constant pressure specific heat of the flow; while the kinetic energy evolution is given by,

$$\frac{\partial}{\partial t} \left(\frac{1}{2} \rho u^2 \right) = -\nabla \cdot \left[\boldsymbol{u} \left(p + \frac{1}{2} \rho u^2 \right) - \boldsymbol{u} \cdot \boldsymbol{S} \right] + \boldsymbol{u} \cdot (\boldsymbol{J} \times \boldsymbol{B}) + \boldsymbol{u} \cdot \boldsymbol{f} - \Phi \tag{5.8}$$

where S is the viscous stress tensor, which is related to the deformation tensor D, by the constitutive equation S = 2νD. The variable ϕ is the vector dissipation term, S:D. The deformation tensor can be expressed as

$$\boldsymbol{D} = \left[D_{ij} \right] = \frac{1}{2} \left[\frac{\partial u_i}{\partial x_j} + \frac{\partial u_j}{\partial x_i} \right] \tag{5.9}$$

We deduce from the latter that due to the action of the Lorentz forces an increase of the kinetic energy leads to a decrease in the magnetic energy. From the temperature equation, Eq. (5.7), one can identify the temporal increase of enthalpy, $\rho c_p \frac{\partial T}{\partial t}$, which equals to the loss of magnetic energy due to joule dissipation, $(1/\sigma)*J^2$, plus the loss of kinetic energy, ϕ, due to viscous dissipation. The ratio of kinetic energy to the accumulated energy can be expressed through the so called *Eckert number*, E_c, which can be expressed numerically as $E_c = \frac{u_0^2}{c_p \Delta T_0}$, where the subscript indicates average or core values. The *Eckert number* could be useful to express dissipation.

From the mathematical point of view, the coupling between Maxwell equations and Navier-Stokes equations induces an additional nonlinearity with respect to the ones already present, leading to unsolved questions of existence and uniqueness (mainly related to the hyperbolic nature of Maxwell equations). As explained by Gerbeau et al., simplified models can be analyzed but care should be taken with certain approximations.

> A system coupling the time dependent incompressible Navier-Stokes equations with a simplified form of the Maxwell equations (low frequency approximation) is well-posed when the electromagnetic equation is taken to be time-dependent, i.e. parabolic form. In contrast, the same model is likely to be ill-posed when the electromagnetic equation is taken to be time-independent, i.e. elliptic form, while the hydrodynamic equations are still in a time dependent form.

A compound of 78 % K and 22 % Na known as NaK78 is the working fluid by excellence for space nuclear reactors. The most common working fluids considered are Na, Li and NaK78. The disadvantage of NaK78 compared to the other options is the fact that it requires more pumping power than Na or Li but when we study other characteristics of these liquid metals we find a series of advantages that NaK78 has over the other elements for space applications. Among the characteristics that lead to the selection of NaK78 as working fluid are:

- NaK78 is liquid at 261 K while Li at 454 K and Na at 361 K;
- Li readily dissolves Nickel at the temperature of interests so it is not suitable to use with steels or super-alloys;
- Neutron capture of ^{6}Li and ^{7}Li produces He gas that would have to be removed from the reactor in space; and
- NaK78 activates less strongly than Na.

The Electric Circuit Design Approach

There are not so many possibilities to design, model and simulate EM pumps and its components. The most widely used method is the electric circuit approach but nowadays modelling using first principles and multiphysics analysis can for first time ever provide a more accurate model. This author, together with a few other specialists in liquid metal engineering MHD, is actively pursuing research activities in the latter and it is expected to have developed a set of tools using first principles in the next few years (Table 5.1).

As a first step, before a complete model can be generated, some main parameters can be obtained by decoupling the electrodynamics from the thermo-fluid component. It should be kept in mind that the parameters obtained in this way are first order approximations; they also constitute the basis for the development of improved models and for a qualitative understanding of the physical phenomena taking place.

Table 5.1 Physical characteristics of Li, Na and Nak78 as working fluids

	Lithium	Sodium	NaK-78
Molecular weight (g/mol)	6.941	22.99	35.55424
MP (°C)	180.54	97.72	−12.6
MP (K)	453,69	370.87	260.55
Operating temperature, T (K)	S50	850	850
ΔT (K)	100	100	100
Density, p (kg/m^3)	476.923	813.912	732.556
Dynamic viscosity, \ (Pa-s)	2.97E-04	2.13E-04	1.90E-04
Surface tension, o$_s$ (N/m)	0.335548	0.150715	0.087088
Thermal conductivity, k (W/m K)	56.465	66.02679	26.02016
Specific heat, Cp (J/K kg)	4165.308	1243.369	875.0363
Electrical conductivity, a$_e$ (1/ Ω m)	2.80E+06	3.21E+06	134E+06
Vapor pressure, P (Pa)	3.993298	2207.764	369358.1
Vapor pressure, P (Torr)	0.03 0351	16.78033	2807.343
Estimated vapor density, p$_g$ (kg/m^3)	3.92E-06	0.007182	1.858164

The Electric Circuit Model

The idea behind the equivalent circuit approach relies on the assumption that the flow is laminar (pressure and velocity independent of time). Hence the electromagnetic and hydrodynamic phenomena can be separated and the theory of linear induction machines and electric circuits can be used, Fig. 5.2. It is assumed that only one phase needs to be considered due to the symmetry a 3-phase balanced system provides. Using this approach, the flow rate, Q, and the developed pressure, ΔP, are connected by the following expression:

$$\Delta P = \frac{3I^2}{Q} \frac{R_2(1-s)}{s(\frac{R_2^2}{X_m^2 s^2} + 1)} \tag{5.10}$$

where X_m is the magnetizing reactance of the pump, R_2 the secondary resistance due to the liquid metal and s is the slip factor ($1 - u_{liq.\ metal}/u_{traveling\ wave}$). A more complicated expression (found by Kim and Hong) with all the pump variables given in an explicit form is:

$$\Delta P = \frac{36\sigma\,sf\,\tau^2(\mu_0 k_w NI)^2}{pg_e^2\left\{\pi^2 + \left(2\mu_0\sigma\,sf\,\tau^2\right)^2\right\}} \tag{5.11}$$

where g_e is the effective inter-core gap coefficient ($K_c{}^*g$, K_c: Carter coefficient and g: inter-core gap), is the pole pitch, p the number of pole pairs, N the number of turns per slot, k_w the winding coefficient, f the input frequency and μ_0 the magnetic permeability of vacuum. As explained by Kim, one can express the resistance of the liquid metal,

$$R_2 = \frac{6\pi D\rho_r'(k_w N)^2}{\tau p} \tag{5.12}$$

as well as the magnetizing reactance of the pump,

$$X_m = \frac{6\mu_0\omega\tau\pi D_0(k_w N)^2}{\pi^2 pg_e}, \tag{5.13}$$

Fig. 5.2 Equivalent electric circuit for an EM pump

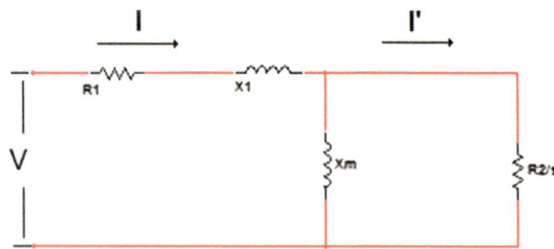

where ρ'_r is the surface resistivity of the fluid (defined as resistivity/fluid thickness), D_0: diameter of the inner core, D: mean diameter of the fluid and g_e: effective inter-core gap (g_e = Carter's coefficient*inter core gap). Unfortunately not all the variables are always known and usually more equations plus a complicated optimization process have to be used.

Another important parameter is the pump efficiency. In a very first approximation, obtained by Araseki et al., it can be expressed as

$$\eta = \frac{\Delta P \cdot Q}{3VIP_f} \tag{5.14}$$

where V is the phase voltage and P_f is the power factor. Unfortunately this expression is not always very useful and it was found by Hee Reyoung Kim that a more useful expression for the type of pump being developed is

$$\varepsilon = \frac{6k_w^2(1-s)}{\frac{\rho_c N_{coils,pole,ph}k_p^2 N_{ph}^2 \sigma g_e}{k_f k_d \tau}\left\{1 + \left(\frac{\pi}{2\mu_0 f s \sigma \tau^2}\right)^2\right\} + \frac{6k_w^2}{s}} \tag{5.15}$$

where k_f: slot filling factor (0.5–0.6), k_d: slot depth/width, N_{ph}: number of input phases and k_p: slot pitch/slot width. In a similar way, an expression linking developed pressure, current, flow rate and length can be derived. Such equation, in principle, would look like

$$\Delta P = \frac{18I^2}{Q} \frac{\pi D \rho'_r \mu_0 f (k_w N)^2 s(1-s)}{3\pi^2 D \rho'_r g_e (k_w N)^2 + \tau^3 \mu_0 p f s^2} \tag{5.16}$$

where $\tau = L/(2p)$. These equations were confirmed experimentally for Na with accuracy between 7 and 17 %, which indicates that other variables might be present. A main contributor to the uncertainties is the so called end effects generated by the distortion of the magnetic fields at both ends of the pump. If the approximation $D \sim D_0$, $g_e \sim g$ and $\mu \sim \mu_o$ is made, the uncertainty can rise as high as 33 %. The length of the pump is given by $L = 2p$ and, due to the existence of teeth and slots, not any pump length is available. Using the Kim-Baker-Tessier approach, the width of the stator slot is given by

$$D_{stator\ slot} = 0.625\frac{\tau}{N_{phases}N_{coils,pole,phase}} \tag{5.17a}$$

where N_{phases} is the number of phases and $N_{coils,\ pole,\ phase}$ is the number of coils per pole per phase. In a similar way, the width of the stator tooth is given by

$$D_{stator\ tooth} = 0.375\frac{\tau}{N_{phases}N_{coils,pole,phase}} \tag{5.17b}$$

and the slot depth by

$$D_{slot\ depth} = 5D_{stator\ slot} \tag{5.18}$$

The clearance between the coil and the end of the stator tooth was estimated to be ~2.54 mm for a voltage of 120 V. The winding distribution factor, k_w, can be expressed as

$$k_w = \frac{\sin(180°/2N_{phases})}{N_{coils,pole,phase}\sin(180°/2N_{phases}N_{coils,pole,phase})} \tag{5.19}$$

For a Na pump similar in design to the one being considered for NaK, it was found at UNIST (by *Hee Reyoung Kim*) and Seoul National University (*Sang Hee Hong*) that the maximum value of the developed force increases with the diameter of the inner core and the efficiency decreases for larger inner core diameters. Also it was found that the lower the input frequency and the longer the core length, the larger the maximum value of the efficiency. The number of turns has no effect on the latter but it should be moderate so as not to require values of the input voltage which are too large due to the increased resistance that results from a large number of turns.

The electromagnetic pumps of the ALIP-type are mainly characterized by three geometric parameters: the length of the core, the diameter of the inner core and the channel gap; as well as by one electrical parameter: the input frequency. An increase in the inter-core gap causes a reduction in the developed force and efficiency. The narrower the gap, the larger the magnetizing reactance, causing an increase of the developed force as well as in efficiency. However, a gap size which is too narrow will produce larger hydraulic frictional loss. The latter is calculated from the Darcy-Weisbach equation,

$$\Delta P_L = \left(fLu^2g/2D\right) \tag{5.20}$$

where f: *friction factor*, L: *length of the flow path*, u: *the flow velocity*, g: *acceleration of gravity*, and D: *hydraulic diameter of the flow channel* $-D_0$ *minus* D_i-.

Three Phase Power System

Thermo-magnetic systems of the ALIP-type make use of a 3-phase alternate current input voltage to power the group of solenoids that generates the traveling magnetic field along the pump duct. The system is assumed to be balanced so only one-phase needs to be considered due to symmetry properties. There are two ways to interconnect the coils: "Y" configuration or "Δ" configuration. The Δ configuration is the most common, and practical, but one has to keep in mind that the values of the equivalent one-phase circuit usually correspond to the equivalent "Y" configuration. A diagram of a "Δ" configuration circuit is shown in Fig. 5.3.

In a balanced system, the currents $I_{aA} = I_{bB} = I_{cC}$; if current measurements indicate otherwise the system is not balanced and that might be an indication of damage or fabrication flaws in the solenoids; I_{aA}, I_{bB}, I_{cC} are known as the line currents, I_{line}; I_{AB}, I_{BC}, I_{CA} are known as the phase currents, I_{phase}. The relationship between currents is $I_{line} = \sqrt{3}I_{phase}$ with a change in phase of 30° (−30° for

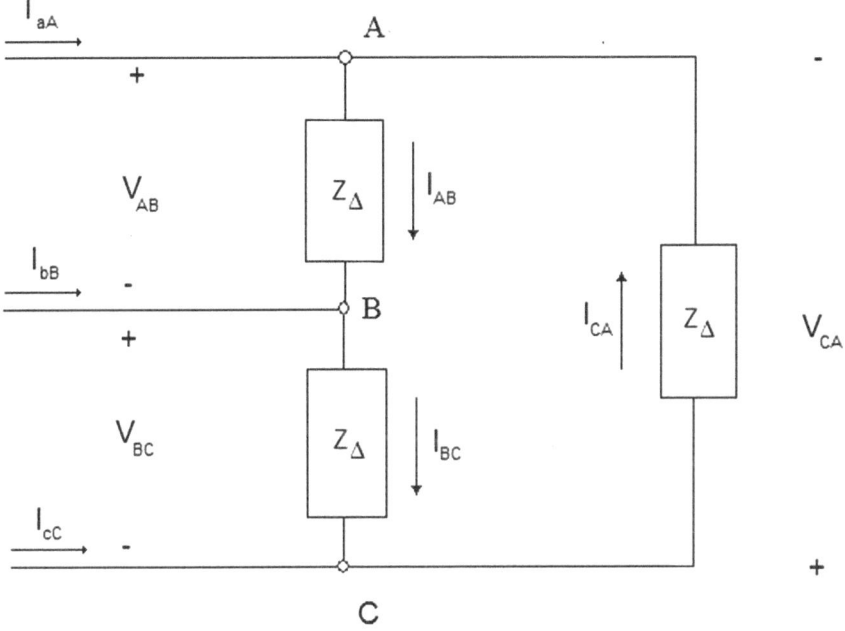

Fig. 5.3 Equivalent circuit diagram for a "Δ" configuration interconnection

a "positive" sequence and $+30°$ for a "negative" phase sequence). In a load connected in "Δ", $V_{line} = V_{phase}$. The power associated to each phase is $P_A = P_B = P_C = P_{phase} = V_{phase}I_{phase}\cos(\theta)$, where θ is the angle between the phase voltage and the phase current. The total power to a balanced "Δ" load would be

$$P_{total} = 3P_{phase} = 3V_{phase}I_{phase}\cos(\theta) = \sqrt{3}V_{line}I_{line}\cos(\theta) \qquad (5.21)$$

For the analysis of the one-phase "Y" equivalent circuit let's recall the equivalent impedance is $Z_Y = (1/3)Z_\Delta$. If the voltage measured is $V_{A\text{-ground}}$ then $V_{AB} = \sqrt{3}V_{A-ground}$ with a phase shift of $30°$. It is important to highlight that knowledge of the phase currents in a "Δ" circuit always uniquely determines the line currents; however, knowledge of the line currents does not allow direct and unique determination of the phase currents for an unbalanced circuit.

A 3-phase power system under balanced operation has the property that the instantaneous power is constant in value. As a consequence, the ALIP or any other device operated from a 3-phase power source exhibit less vibration and noise than a mono-phase device. Measurement of the average power flow is done with a watt-meter. The wattmeter has a voltage coil and a current coil and it reads the average value of power flowing into a single-phase network as

$$P = \frac{1}{T}\int_0^T v(t)i(t)dt \qquad (5.22)$$

and the apparent power flowing to the network, if both v and i are void or harmonics, is given by $S = VI$ where V and I are the *rms* values of $v(t)$ and $i(t)$. The power factor follows as $PF = P/S$. Measurement of average power flow in a 3-phase circuit can be done by connecting a wattmeter to read the power flowing to each phase and adding the three readings. However, if the network is "Y" connected, the potential coils of the three watt-meters is required to be connected to the neutral which usually is not available in a balanced system. If the system is "Δ" connected, each branch of the "Δ" must be opened to insert the wattmeter current coils which it is often not feasible. An approach to solve the problems just described is by making use of the fact that the total 3-phase average power can be written as

$$P_T = V_{BC}I_{bB}\cos\left(\theta_{V,BC} - \theta_{i,bB}\right) + V_{AC}I_{aA}\cos(\theta_{V,AC} - \theta_{aA}) \qquad (5.23)$$

so two watt-meters, connected as shown in Fig. 5.4, can be used to measure PT. The total 3-phase average power is the algebraic sum of the readings whether the system is balanced or unbalanced.

The power source of the ALIP should be able to generate a sine wave signal to drive the solenoids; if the power source cannot generate a high fidelity sine wave then a sine wave filter should be used. Since nonlinear components generate harmonic currents, any distortion of an originally pure sine wave constitutes harmonic frequencies. When the sine-wave distortion is symmetrical above and below

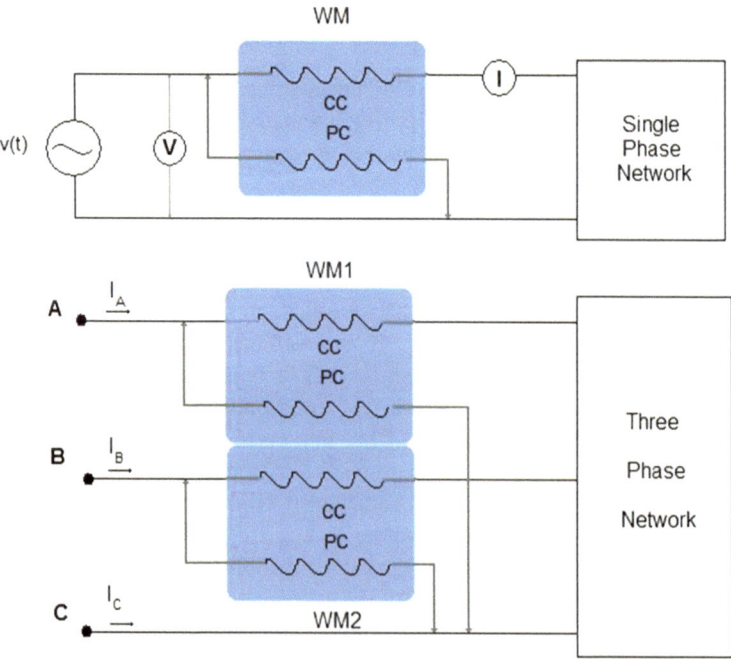

Fig. 5.4 Power measurement using a wattmeter. *Top*, single phase application. *Bottom*, two watt-meter method for 3-phase networks

the average centerline of the waveform, the only harmonics present will be odd-numbered, not even-numbered. The 3rd harmonic, and integer multiples of it, are known as triplen and they are in phase with each other, despite the fact that their respective fundamental waveforms are 120° out of phase with each other. Triplen harmonic currents in a "Δ" connected set of components circulate within the loop formed by the "Δ".

The three-phase winding arrangement for the solenoids usually follows the sequence AA ZZ BB XX CC YY where A, B, C denote the balanced three-phase winding and X, Y, Z the opposite phase; for a direct balanced system, if A: 0°, B: 120° and C: 240° then X: 180°, Y: 300° and Z: 60°. Arranging the sequence by rising phase, one obtain the correct winding sequence for the solenoids: AA ZZ BB XX CC YY (as seen in Fig. 5.5). Sometimes a phase shift to this arrangement could minimize certain instabilities present in the pump but a lower efficiency of the ALIP is the cost to pay for such procedure.

Fig. 5.5 Solenoid phase distribution diagram where A, B, C is the balance 3-phase winding and X, Y, Z is the opposite 3-phase winding. The *bottom* figure shows the peak current density induced in the pump walls in a specific time period

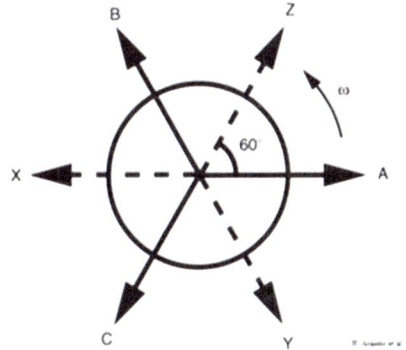

Solenoidal Fields and End Effects

Solenoids are electromagnetic elements that can be used in many different applications in many different ways. Besides its useful characteristics, their fringe
fields propagate over long distances and even though this could be beneficial
in certain situations, care must be taken when including them in any design.
When a charged lagrangian particle crosses a solenoid fringe fields, it experiences a transverse force causing the particle to follow a spiral path and a coupling between coordinates. Fringe fields are used as synonymous of edge fields,
representing the boundary region between the field and no field region with a
quick fall off. These fringe fields generated by coils contain various nonlinearities due to the longitudinal dependence of the field. In practice it is important to
be able to efficiently combine these fields consisting of several solenoidal coils.
This could simplify the simulation efforts due to the shortened fringe fields created by the field cancellation of counteracting coils.

A field distribution for thick solenoids, in agreement (7 % uncertainty) with
the coils used in the ALIPs (Fig. 5.6), is given by the so called CMST function
which describes a thick solenoid extending longitudinally from $s = 0$ to $s = l$,

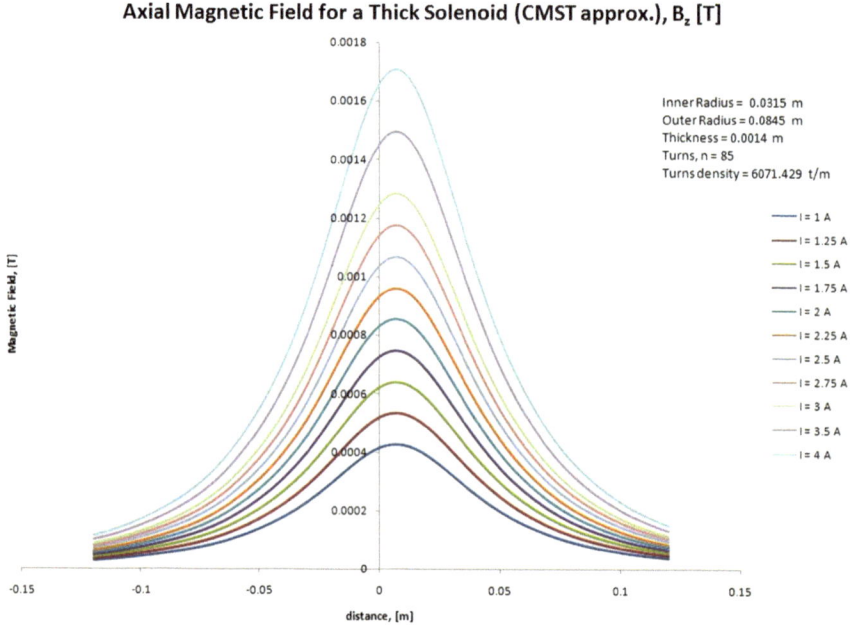

Fig. 5.6 Axial magnetic field for the ALIP solenoids when a DC current is applied. Note the
extension of the fringe fields. When an AC current is used instead of a DC signal, a sinusoidal component should be added and, in occasions, an additional term to compensate for the frequency response

and radially from $r = R_1$ to $r = R_2$, with a winding density n and carrying a current I.

$$B_z, \text{CMST}(s) = \frac{\mu_0 I n}{2(R_2 - R_1)} \left[s \log\left(\frac{R_2 + \sqrt{R_2^2}}{R_1 + \sqrt{R_1^2 + s^2}}\right) - (s - l) \log\left(\frac{R_2 + \sqrt{R_2^2 + (s-l)^2}}{R_1 + \sqrt{R_1^2 + (s-l)^2}}\right) \right]$$

(5.24)

Experimentally, once obtained the axial component of the magnetic field, their values in other regions of space can be determined by the application of Maxwell equations assuming no coil assembly flaws. For higher precision needs, a computer controlled 3D scanning probe should be used.

Due to the solenoids fringe fields and its finite core length, the ALIP has an *end effect* at both ends of the pump—where the magnetic field is distorted-. A reduction on the developed force arises which is roughly equal to the product of the magnetic field and its perpendicular induced current. Theoretical calculations indicate a reduction of the end effects by controlling the input frequency; increasing the efficiency at the lower frequencies compared with results obtained at frequencies over 60 Hz. The inlet *end effect* force affects most of the pump while the outlet *end effect* domain is limited to the exit region. Considering the direct relationship between fluid velocity and *end effect*, low frequency operation is preferred as far as the developing force and the efficiency are not decreased too much. When the end effects are neglected, it is easy to show that the pump efficiency is given as the ratio of the flow velocity u to the synchronous velocity/of the fields (i.e. n = κu/ = $1-s$). But when the end effects are included the pump efficiency has to be computed using numerical integration. Note that the maximum pump efficiency for ALIPs lies in the range $0.2 < s < 0.4$ in many cases (Fig. 5.7).

One last comment regarding the skin effect of the induced currents: the width of the fluid channel in the ALIPs and other types of electromagnetic pumps should be limited to below the skin depth of the fluid for stable operation.

$$skin\,depth = \delta = \frac{1}{\sqrt{\pi f u \sigma}}$$

(5.25)

Fig. 5.7 Comparison of pump efficiencies between two cases with and without end effects when the pump operates with different flow velocities expressed in terms of the slip. *Courtesy* Cho and Hong (1998)

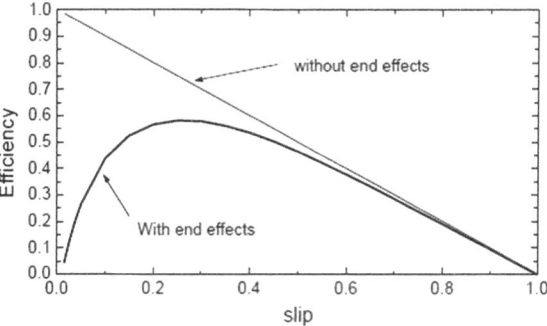

Instabilities and Flow

A variation of the potential generated by the solenoids along the perimeter of the duct drive currents along the conducting walls and inside the viscous layers near them. The components of current which are perpendicular to the magnetic field induce a Lorentz force. The Lorentz force contributes to the pressure drop in addition to the viscous pressure drop in hydrodynamic drop. The electromagnetic pressure drop may even exceed the viscous pressure drop by orders of magnitude for strong magnetic fields.

Liquid metals have typically high electrical conductivity giving rise to large Hartmann numbers of the order of $\sim 10^3$–10^4 for moderate or strong magnetic fields. In this case, viscous effects are confined to the near vicinity of the walls for homogenous magnetic fields and inertia effects become smalls for high inter-action parameters. The theory of Hartmann flow cannot always being applied for time dependent magnetic fields and further studies should be carry on. For high Ha numbers usually free shear layers originate from corners in duct expansions. These shear layers are in the direction of the external magnetic field and may give rise to a significant flow redistribution and contribute to the pressure drop. The sudden occurrence of a vertical velocity component in the expanding section gives rise to a local shear flow. In the case of a constant magnetic field, expansions in the direction of the field cause maximum 3D pressure drop, while the extra pressure drop caused by expansions in the plane perpendicular to the field is negligible. Bends and expansions that turn the flow into the direction of the magnetic field should be avoided in engineering applications in order to minimize pressure drop or to exclude undesirable flow distributions. For the case of time dependent magnetic fields a solution could be to calculate the average field component and its gradient; trying to make the expansion to follow the gradient direction to minimize pressure drops or flow redistribution if possible.

As explained before, it was found experimentally that instabilities arise when the following three conditions hold true:

- $R_{ms} > 1$,
- $D/2\tau =$ "*large enough*"
- $N_{int} =$ "*large enough*"

where $D/2\tau =$ ratio of the mean radius to the pole pitch and $N_{int} =$ modified interaction parameter (also known as modified Stuart number). The modified interaction parameter is expressed as $N_{int} = 2gB_{ar}^2\sigma/\rho\lambda v_b$ where B_{ar} is the radial component of the applied magnetic field and λ the friction coefficient. This specific instability appears as a low frequency (LF) pulsation in the pressure head affecting the flow rate, liquid metal velocity and magnetic field distribution. As a consequence vortices are generated in the inlet region as well as fluctuations in the winding currents and voltages. The dominant frequency of this instability is in the range 0–10 Hz with amplitude that increases with the slip factor. This LF pressure pulsation is produced by vortices in the liquid metal generated by the no uniform-ity of the azimuthal component of the applied magnetic field when $R_{ms} > 1$.

The magnetic Reynolds number, R_{ms}, often becomes greater than unity in the pump region where the slip is larger than 0.2 giving place to another instability known as double supply frequency (DSF) pressure pulsation. Experimental work done by Araseki et al. shows that only the vibration caused by the DSF pressure pulsation occurs in the pump outlet and propagates to the pipe when $R_{ms} < 1$ and $s < 0.2$. In the case of multiple pole pair pumps, this instability could be reduced by shifting the phase angle between pole and pole (i.e. $60°$ phase shift) but lowering the efficiency as a result of such modification.

Bibliography

Cho, S., Hong, S.H.: The magnetic field and performance calculations for an electromagnetic pump of a liquid metal. J. Phys. D: Appl. Phys. 31 (1998)

Borghi, C.A. et al.: Study of the design model of a liquid metal induction pump. IEEE Trans. Magn. **34**(5) (1998)

Araseki, H. et al.: Magnetohydrodynamic instability in annular linear induction pump. Part I & II. Experiment and numerical analysis. Nucl. Eng. Des. **227**, Elsevier (2004)

Gerbeau, J. et al.: Mathematical methods for the magnetohydrodynamics of liquid metals. Oxford University Press, Oxford (2006)

Muller, U., Buhler, L.: Magnetohydrodynamics on channels and containers. Springer, Berlin (2001)

Kim, H.R.: Design and experimental characterization of an EM pump, J. Korean Phys. Soc. **35**(4) (1999)

Kim, H.R., et al.: Development and verification of design code for small annular linear induction EM pump, the korean nuclear society spring meeting. Pohang University, Pohang (1999)

Texas Instruments, Technical notes for the design of high frequency transformers (1996)

Baker, R.S., Tessier, M.J.: Handbook of electromagnetic pump technology. Elsevier, Amsterdam (1987)

Maidana, C.O.: Design of a cabinet safe system for a portable particle accelerator, VDM Verlag (2009)

Makino, K., Berz, M.: Solenoid elements in COSY INFINITY, Institute of Physics Conference Series N **175** (2004)

Maidana, C.O. et al.: Design of an annular linear induction pump for nuclear space applications. Nucl. Emerg. Technol. Space Explor. (2011) (NETS2011)

Chapter 6
Assembly and Fabrication Considerations of Annular Linear Induction Pumps for Space Nuclear Reactors

Abstract In the design process of liquid metal thermo-magnetic systems in general and of annular linear induction pumps for space applications in particular, special care should be taken with the materials selected, its fabrication and assembly methodology, instrumentation and thermal stresses. The later requires not only a study of the topics mentioned but also an exhaustive quality assurance and quality control mechanism to be in place. The minimum requirements for these systems are 5–7 years with no fail of any type and a lifespan of at least 8 years for fission surface power applications. If the thermo-magnetic system will be part of a nuclear propulsion system its life span and mean time between failures could be increased even more. Obviously secondary/redundancy mechanisms will be in place but each component added will add weight and add an extra element to the risk management consideration. We need to keep clear that maintenance in space is almost impossible and that liquid metals such as NaK78 are reactive to air and water. In this chapter some considerations to the assembly and fabrication are given.

Keywords Liquid metals · Test loop · ALIP fabrication · Electromagnetic pump · Thermo-magnetic system · Thermal control · Engineering MHD · Assembly

Basic Materials and Heat Transfer Considerations

The most suitable material for the ALIP ducts and outer shell is 316 stainless steel. Grade 316 is the standard molybdenum-bearing grade, which gives it a better overall corrosion resistant property than grade 304. The austenitic structure gives it good toughness properties in a wide range of temperature. Its density is 8,000 kg/m^3, its thermal conductivity is 21.5 W/m K at 500 °C and its tensile strength is between 500 and 600 MPa.

The inner core and stator material could be laminated silicon steel which has anisotropic heat conductivity, Fig. 6.1. The heat conductivity in the lamination plane

© The Author(s) 2014
C.O. Maidana, *Thermo-Magnetic Systems for Space Nuclear Reactors*,
SpringerBriefs in Applied Sciences and Technology,
DOI 10.1007/978-3-319-09030-6_6

Fig. 6.1 Simplified model of an ALIP cross section showing the torpedo with *inner core* structure

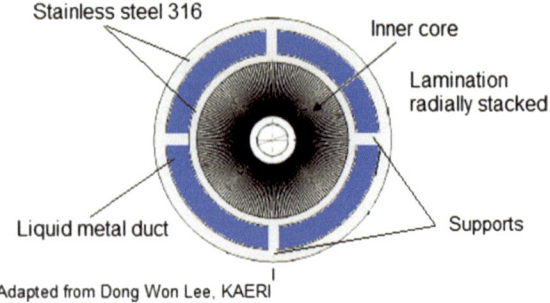

Adapted from Dong Won Lee, KAERI

is between 5 and 30 times higher than the component in the stacking direction. The laminated material reduces the losses produced by the induced eddy currents. For acceptable losses, flux density swing ΔB must be restricted to a value much less than the magnetic flux density at the material saturation limit, B_{sat}.

$$\Delta B = \frac{1}{N \cdot A_e} \int E \, dt \qquad (6.1)$$

where N is the number of turns, A_e the core cross section and E dt is the applied voltage per second. We must emphasis that ΔB is the total peak flux swing whereas $\Delta B/2$ should be applied in the core loss curves.

The solenoids are primarily made of Cu; however, different fabrication processes should be taken into consideration for a more accurate modeling. Besides the thermal conductivity terms, the power loss due to the circulating current should be considered in the heat transfer analysis. One option for the coil conductors is alumina-dispersed-strengthened copper which has a small volume expansion coefficient. An option found in the commercial market and described by *Baker and Tessier* is Cu strip conductors with HML coating plus double glass insulation, silicon impregnated and one layer silicon bonded mica tape all around the coil. The material selected by this author for the U.S. Department of Energy/NASA's Fission Surface Power Technology (FSP) Unit and advised for future space applications is glidcop AL-15, which is a low alumina content grade of dispersion strengthened copper consisting of a pure copper matrix containing finely dispersed sub-microscopic particles of Al_2O_3 which act as a barrier to dislocation movement. The dispersed Al_2O_3 is thermally stable so that it acts to retard recrystallization of the copper. Consequently, significant softening does not occur as the result of high temperature exposure. Along with superior strength retention, thermal and electrical conductivities are higher than conventional copper alloys.

The average thermal conductivity coefficient for NaK78 is 25.3 W/K m. Since the NaK thermal conductivity is so high, the principal mode of energy transfer is conduction, and not convection as it could be firstly assumed. A layer with thermal insulation material can be placed around the external liquid metal duct wall to increase the heat flow in the duct for non-space applications but the lack of radiative/convective heat transfer in space leads to the decision of not to follow this method to ease the heat flux from the solenoids to the duct; in this way the liquid metal is used to dissipate the heat from the stator-solenoid region. There are many materials available with thermal conductivities ranging from 0.1 to 1 W/K m.

The outer shell can be filled using a Helium gas mixture, 80 % He and 20 % Ar, at 400 K inlet temperature between the stator and the external shell to minimize thermal shocks. Helium can also be used in the gap between the outer core and the stator-solenoids for non-space applications. Another material to consider is ceramic aluminum nitrate which is not only a good electrical insulator but a good thermal conductor as well. A material like aluminum nitride can be used to fill the space between stator, coils and outer shell but cracks in the material must be expected. Such cracks are not expected to cause serious problems because they will be contained by the outer shell and the different internal components of the ALIP. Ceralloy 1370-DP is a good option of ceramic aluminum nitrate.

A 3D model can be seen in Fig. 6.2. The inner core, or torpedo, extends beyond the solenoid-stator section due to the fact that the solenoid fringe fields extend over long distances and generating instabilities on the fluid. Experimental data have established a limit to the temporal variation in temperature that an ALIP-type electromagnetic pump can safely handle without structural damage due to thermal effects. The different elements should not be heated at a rate larger than 1–16 °C per minute depending of the materials selected and the

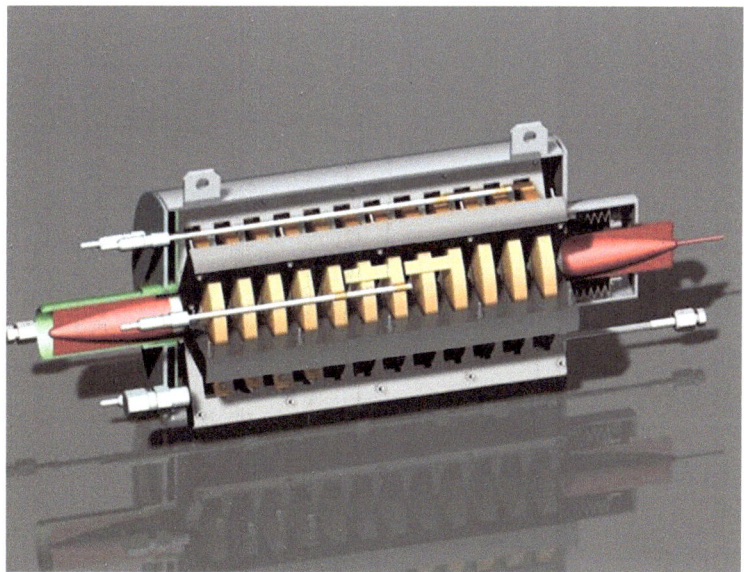

Fig. 6.2 A 3D model of an ALIP-type electromagnetic pump where the coils, stator, torpedo, instrumentation and external shell can be easily appreciated. The inner core (or torpedo) extends beyond the solenoid-stator section due to the solenoid fringe fields that extend over long distances generating instabilities on the fluid

fabrication processes. The thickness of the duct walls affect the efficiency of the pump due to the magnetic resistivity that presents. To avoid possible structural failures a system of three stainless steel rings could be placed around the pump outer shell as well.

Test Loops and Instrumentation

From the research point of view, different MHD devices should be built and tested to better understand its engineering MHD: theory, modeling process, fabrication methods. From the design and manufacturing point of view, one has to build and test a prototype device before the final design is sent for production and assembly. Whatever the case is: research or technical engineering, the prototype has to be tested in a liquid metal test loop. Care has to be taken for safety reasons and to get the right measurements out of the test. Sensors, or gauges, should be positioned in pairs, at 90° one of the each other and they should not be located, if possible, at the exit or entrance sections of the torpedo due to the many instabilities that exists. The fact that the sensors in contact with the liquid metal should come in pairs and at 90° of the each other is exactly to avoid false readings caused by local instabilities. If high resolution measurements are needed, ultrasound or imaging, are the best options (Fig. 6.3).

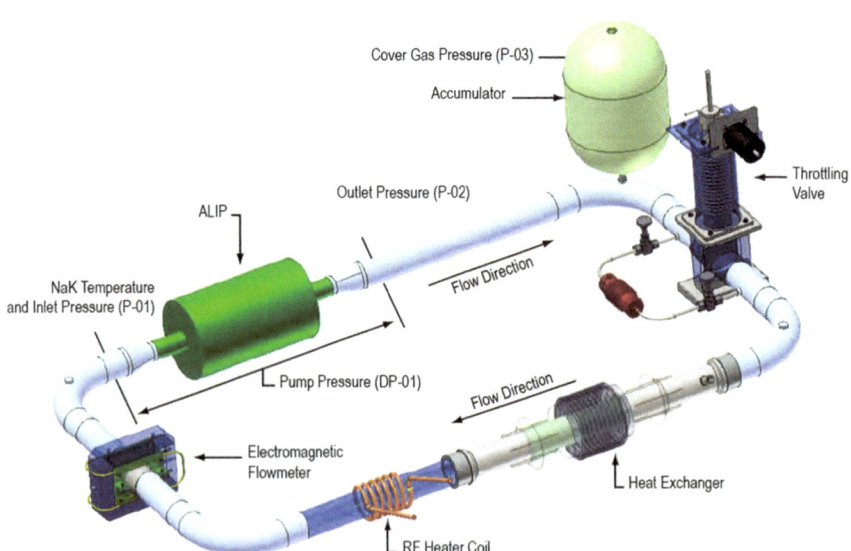

Fig. 6.3 Experimental liquid metal test loop

Bibliography

Maidana, C.O., et al.: Design of an annular linear induction pump for nuclear space applications. In: Proceedings of Nuclear and Emerging Technologies for Space Exploration 2011 (NETS2011), Albuquerque, NM, 7–10 Feb 2011

Werner, J., et al.: An overview of facilities and capabilities to support the development of nuclear thermal propulsion. In: Proceedings of Nuclear and Emerging Technologies for Space Exploration 2011 (NETS2011), Albuquerque, NM, 7–10 Feb 2011